Recombinant DNA Techniques

A Textbook

Recombinant DNA Techniques

A Textbook

Monika Jain

Alpha Science International Ltd.
Oxford, U.K.

Recombinant DNA Techniques
A Textbook
288 pgs. | 65 figs.

Monika Jain
Department of Biotechnology
Sharda University
Greater Noida, UP

ALPHA SCIENCE INTERNATIONAL LTD.
7200 The Quorum, Oxford Business Park North
Garsington Road, Oxford OX4 2JZ, U.K.

www.alphasci.com

ISBN 978-1-84265-667-9

Printed in India

Acknowledgement

It is an honor for me to thank those who made this book possible. I owe my deepest gratitude to my loving husband, Devendra, whose encouragement, guidance and support from the initial to final level enabled me to develop this book. I would also like to thank my parents, Mr. Rajendra Kumar Jain and Smt. Sharda Jain, who always encouraged me to write this book. I would also like to show my gratitude to my brother, Rohit, who helped me in number of ways.

Lastly I offer my regards to all of those who supported me in any respect during the completion of the book.

Monika Jain

Preface

Recombinant DNA technique is a technique which allows DNA to be produced via artificial means. The procedure has been used to change DNA in living organisms and may have even more practical uses in the future. It is an area of medical science that is just beginning to be researched in a concerted effort. Recombinant DNA technique works by taking DNA from two different sources and combining that DNA into a single molecule. That alone, however, will not do much. Recombinant DNA technique only becomes useful when that artificially-created DNA is reproduced. This is known as DNA cloning. There are two main types of cloning that recombinant DNA technique is used for: therapeutic cloning and reproductive cloning. Most are familiar with reproductive cloning, which will produce an organism with the exact genetic information of an already-existing organism. This has already been done with some animals. Dolly, a sheep, was the first mammal to ever be reproduced as an exact genetic copy. Therapeutic cloning is cloning that is used to reproduce certain tissues or organs, not an entire organism.

Using recombinant DNA technique for therapeutic cloning purposes holds a great deal of potential benefit. For example, a cancerous organ could be replaced with a new one made from a patient's own DNA. This would likely help reduce the rejection of organs that sometimes happens when a transplant takes place. If a heart is damaged, it could even be replicated using recombinant DNA technique. While these applications may be years away from practical use, they are possibilities. In addition, there are a number of other uses for recombinant DNA technique. It may help make crops more resistant to heat and drought. It may even be used to create plans with genes that repel harmful insects. In such cases, it would be like the plant had a built-in insecticide or repellent, reducing the need for humans to handle harmful chemicals. Recombinant DNA technique is not accepted among some, especially social conservatives, who feel the technology is a slippery slope to devaluing the uniqueness of life. Further, because some DNA work involves the use and destruction of embryos, there is more controversy created. Still, proponents of the technique say the ultimate goal is to benefit human life, not destroy it.

Recombinant DNA technique is one of the most exciting areas of Biotechnology and its applications have to reach the common man. It has attracted worldwide attention and is considered a field that can end the world's hunger in a lesser duration. It is no longer a specialized segment of science confined to high tech laboratories. The subject of Recombinant DNA Technique is now entering into degree and even pre-university classes.

The "Recombinant DNA Techniques" has been written to meet the requirements of the students of B.Sc., M.Sc., B.Tech. in Biotechnology of various universities of India. It provides an insight into the basics of the recombinant DNA technology and genetic engineering. It covers the basic information about the PCR, gene sequencing, cDNA cloning, various DNA transformation methods, basic idea about the molecular markers and gene therapy.

Though I have tried to make the book student friendly and comprehensive in all respects, it is possible that some errors might still have been left due to oversight. Any suggestions or constructive criticism from the readers will be highly appreciated and considered.

Monika Jain

Contents

1

Introduction to Recombinant DNA Technology

Recombinant DNA is a form of DNA that does not exist naturally, which is created by combining DNA sequences that would not normally occur together. In terms of genetic modification, recombinant DNA (rDNA) is introduced through the addition of relevant DNA into an existing organismal DNA, such as the plasmids of bacteria, to code for or alter different traits for a specific purpose, such as antibiotic resistance. It differs from genetic recombination, in that it does not occur through processes within the cell, but is engineered. A recombinant protein is a protein that is derived from recombinant DNA.

Recombinant DNA technology is a technology which allows DNA to be produced via artificial means. The procedure has been used to change DNA in living organisms and may have even more practical uses in the future.

Recombinant DNA technology refers to the set of techniques for recombining genes from different sources *in vitro* and transferring this recombinant DNA into cells where it may be expressed. These techniques were first developed around 1975 for basic research in bacterial molecular biology, but this technology has also lead to many important discoveries in basic eukaryotic molecular biology. Such discoveries resulted in the appearance of the biotechnology industry. Biotechnology refers to the use of living organisms or their components to do practical tasks such as:

The use of microorganisms to make wine and cheese
Selective breeding of livestock and crops
Production of antibiotics from microorganisms
Production of monoclonal antibodies

The use of recombinant DNA techniques allows modern biotechnology to be a more precise and systematic process than earlier research methods. It is also a powerful tool since it allows genes to be moved across the species barrier. Using these techniques, scientists have advanced our understanding of eukaryotic molecular biology.

The Human Genome Project is an important application of this technology. This project's goal is to transcribe and translate the entire human genome in order to better understand the human organism. A variety of applications are possible for this technology, and the practical goal is the improvement of human health and food production.

Recombinant DNA technology works by taking DNA from two different sources and combining that DNA into a single molecule. That alone, however, will not do much. Recombinant DNA technology only becomes useful when that artificially-created DNA is reproduced. This is known as **DNA cloning**.

There are two main types of cloning that recombinant DNA technology is used for: **therapeutic cloning and reproductive cloning**. Most are familiar with reproductive cloning, which will produce an organism with the exact genetic information of an already-existing organism. This has already been done with some animals. Dolly, a sheep, was the first mammal to ever be reproduced as an exact genetic copy. Therapeutic cloning is cloning that is used to reproduce certain tissues or organs, not an entire organism.

Using recombinant DNA technology for therapeutic cloning purposes holds a great deal of potential benefit. For example, a cancerous organ could be replaced with a new one made from a patient's own DNA. This would likely help reduce the rejection of organs that sometimes happens when a transplant takes place. If a heart is damaged, it could even be replicated using recombinant DNA technology. While these applications may be years away from practical use, they are possibilities.

In addition, there are a number of other uses for recombinant DNA technology. It may help make crops more resistant to heat and drought. It may even be used to create plants with genes that repel harmful insects. In such cases, it would be like the plant had a built-in insecticide or repellent, reducing the need for humans to handle harmful chemicals.

Currently, recombinant DNA technology has attracted headlines when it has been used on animals, both to create identical copies of the same animal or to create entirely new species. One of those new species is the GloFish™, a type of fish that seems to glow with a bright fluorescent coloring. While they have become a popular aquarium fish, they have other uses as well. Scientists hope to use them to help detect polluted waterways, for example.

OVERVIEW OF THE RDT TECHNIQUES

DNA Cloning

DNA technology makes it possible to clone genes for basic research and commercial applications. Prior to the discovery of recombinant DNA techniques, procedures for altering the genes of organisms were constrained by the need to find and propagate desirable mutants. Geneticists relied on natural processes, mutagenic radiation, or chemicals to induce mutations. In a laborious process, each organism's phenotype was checked to determine the presence of the desired mutation. Microbial geneticists developed techniques for screening mutants. For example bacteria were cultured on media containing an antibiotic to isolate mutants which were antibiotic resistant **(Fig. 1.1).**

Before 1975, transferring genes between organisms was accomplished by cumbersome and nonspecific breeding procedures. The only exception to this was the use of bacteria and their phages.

a) Genes can be transferred from one bacterial strain to another by the natural processes of transformation, conjugation and transduction.
b) Geneticists used these processes to carry out detailed molecular studies on the structure and functioning of prokaryotic and phage genes.
c) Bacteria and phages are ideal for laboratory experiments because they are relatively small, have simple genomes, and are easily propagated.
d) Although the technique was available to grow plant and animals cells in culture, the workings of their genomes could not be examined using existing methods.

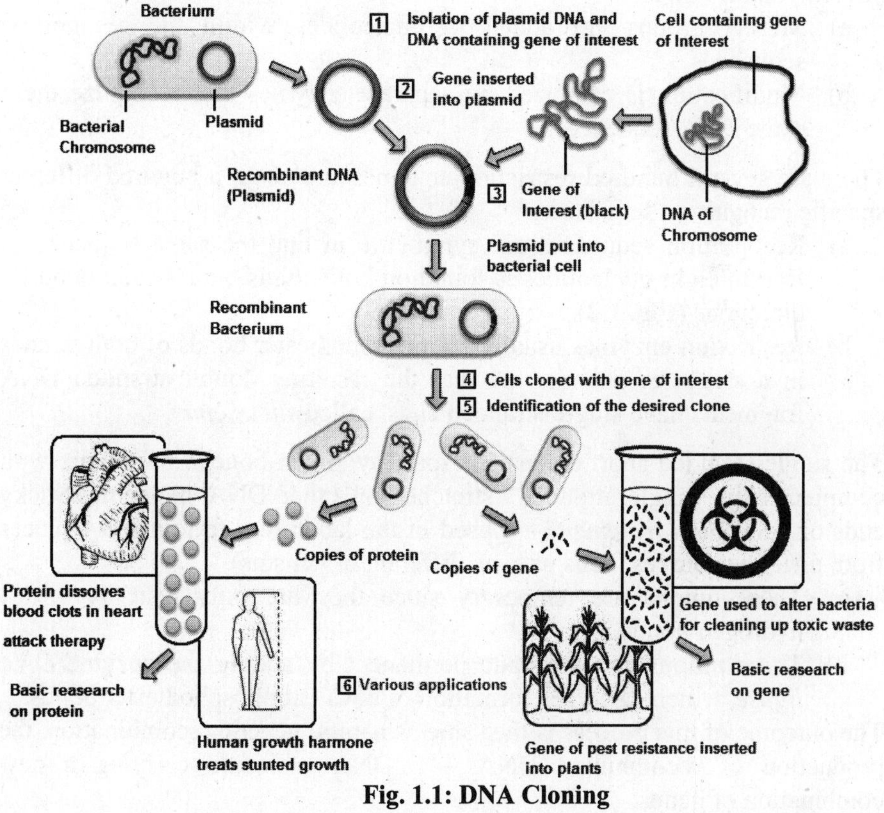

Fig. 1.1: DNA Cloning

1.4 *Recombinant DNA Techniques*

Recombinant DNA technology now makes it possible for scientists to examine the structure and function of the eukaryotic genome, because it contains several key components:

a) Biochemical tools that allow construction of recombinant DNA.
b) Methods for purifying DNA molecules and proteins of interest.
c) Vectors for carrying recombinant DNA into cells and replicating it
d) Techniques for determining nucleotide sequences of DNA molecules.

Restriction enzymes are used to make recombinant DNA

Restriction enzymes are major tools in recombinant DNA technology.

a) First discovered in the late 1960s, these enzymes occur naturally in bacteria where they protect the bacterium against intruding DNA from other organisms.
b) This protection involves, *restriction*, a process in which the foreign DNA is cut into small segments
c) Most restriction enzymes only recognize short, specific nucleotide sequences called *recognition sequences* or restriction sites. They only cut at specific points within those sequences.

Bacterial cells protect their own DNA from restriction through *modification* or methylation of DNA.

a) Methyl groups are added to nucleotides within the recognition sequences.
b) Modification is catalyzed by separate enzymes that recognize these same DNA sequences.

There are several hundred restriction enzymes and about a hundred different specific recognition sequences.

a) Recognition sequences are symmetric in that the same sequence of four to eight nucleotides is found on both strands, but run in opposite directions **(Fig. 1.2)**.
b) Restriction enzymes usually cut phosphodiester bonds of both strands in a staggered manner, so that the resulting double-stranded DNA fragments have single-stranded ends, called *sticky ends*

The single-stranded short extensions form hydrogen-bonded base pairs with complementary single stranded stretches on other DNA molecules.Sticky ends of *restriction fragments* are used in the laboratory to join DNA pieces from different sources (cells or even different organisms).

a) These unions are temporary since they are only held by a few hydrogen bonds.
b) These unions can be made permanent by adding the enzyme *DNA ligase*, which catalyzes formation of covalent phosphodiester bonds.

The outcome of this process is the same as natural genetic recombination, the production of recombinant DNA – a DNA molecule carrying a new combination of genes.

Genes can be cloned in recombinant DNA vectors

Most DNA technology procedures use carriers or vectors for moving DNA from test tubes back into cells.

Cloning vectors – A DNA molecule that can carry foreign DNA into a cell and replicate there:

a) Two most often used types of vectors are bacterial plasmids and viruses.
b) Restriction fragments of foreign DNA can be spliced into bacterial plasmid without interfering with its ability to replicate within the bacterial cell. Isolated recombinant plasmids can be introduced into bacterial cells by transformation.

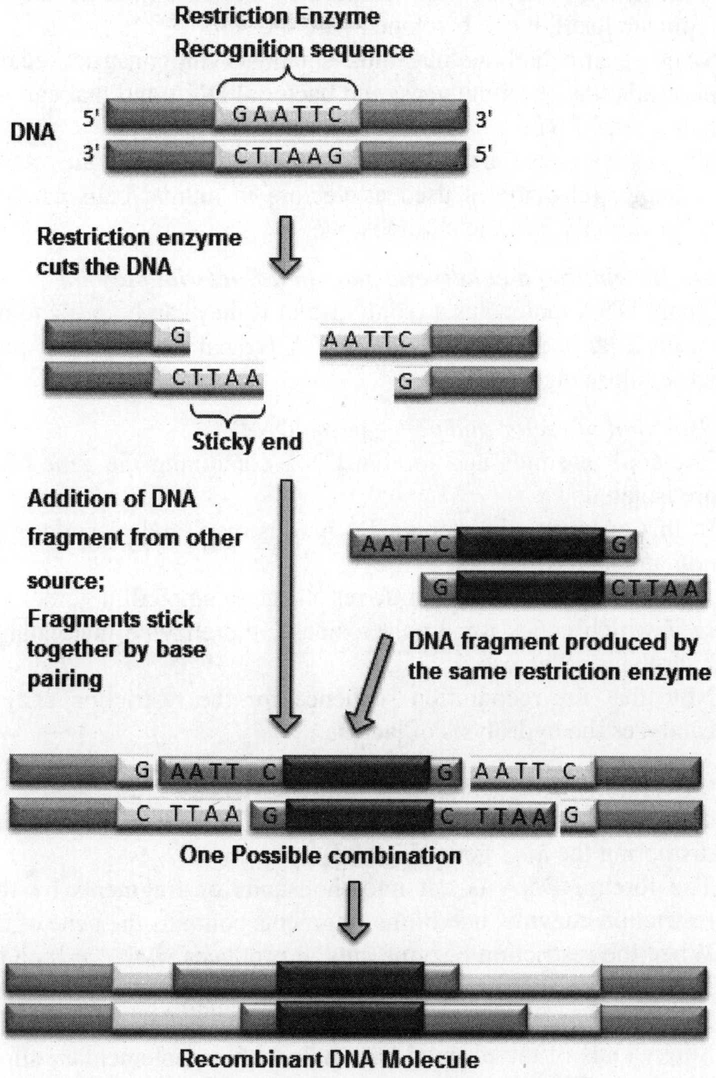

Fig. 1.2: Production of Recombinant DNA

Bacteriophages, such as lambda phage, can also be used as vectors.

a) The middle of the linear genome, which contains nonessential genes, is deleted by using restriction enzymes.
b) Restriction fragments of foreign DNA are then inserted to replace the deleted area.
c) The recombinant phage DNA is introduced into an *E.coli* cell.
d) The phage replicates itself inside the bacterial cell.
e) Each new phage particle carries the foreign DNA "passenger."

Sometimes it is necessary to clone DNA in eukaryotic cells rather than in bacterial. Under the right conditions, yeast and animal cells growing in culture can also take up foreign DNA from the medium.

a) If the new DNA becomes incorporated into chromosomal DNA or can replicate itself, it can be cloned with the cell.
b) Since yeast cells have plasmids, scientists can construct recombinant plasmids that combine yeast and bacterial DNA and that can replicate in either cell type.
c) Viruses can also be used as vectors with eukaryotic cells. For example, retroviruses used as vectors in animal cells can integrate DNA directly into the chromosome.

Procedure for cloning a eukaryotic gene in a Bacterial plasmid

Recombinant DNA molecules are only useful if they can be made to replicate and produce a large number of copies. A typical gene-cloning procedure includes the following steps:

Step 1: *Isolation of vector and gene-source DNA*

a) Bacterial plasmids and foreign DNA containing the gene of interest are isolated
b) In this example, the foreign DNA is human, and plasmid is from *E. coli* and has two genes:
c) amp^R which confers antibiotic resistance to ampicillin
d) *lacZ* which codes for ß-galactosidase, the enzyme that catalyzes the hydrolysis of lactose
e) Note that the recognition sequence for the restriction enzyme that catalyzes the hydrolysis of lactose

Step 2: *Insertion of gene-source DNA into the vector*

a) The restriction enzyme cuts plasmid DNA at the *restriction site*, disrupting the *lacZ* gene **(Fig. 1.3).**
b) The foreign DNA is cut into thousands of fragments by the same restriction enzyme; one of the fragments contains the gene of interest.
c) When the restriction enzyme cuts, it produces *sticky ends* on both the foreign DNA fragments and the plasmid
d) Mixture of foreign DNA fragments with clipped plasmids
e) Sticky ends of the plasmid base pair with complementary sticky ends of foreign DNA fragments.

f) Addition of DNA ligase
g) DNA ligase catalyzes the formation of covalent bonds, joining the two DNA molecules and forming a new plasmid with recombinant DNA

***Step 3**: Introduction of cloning vector into bacterial cells*

a) The naked DNA is added to a bacterial culture.
b) Some bacteria will take up the plasmid DNA by transformation

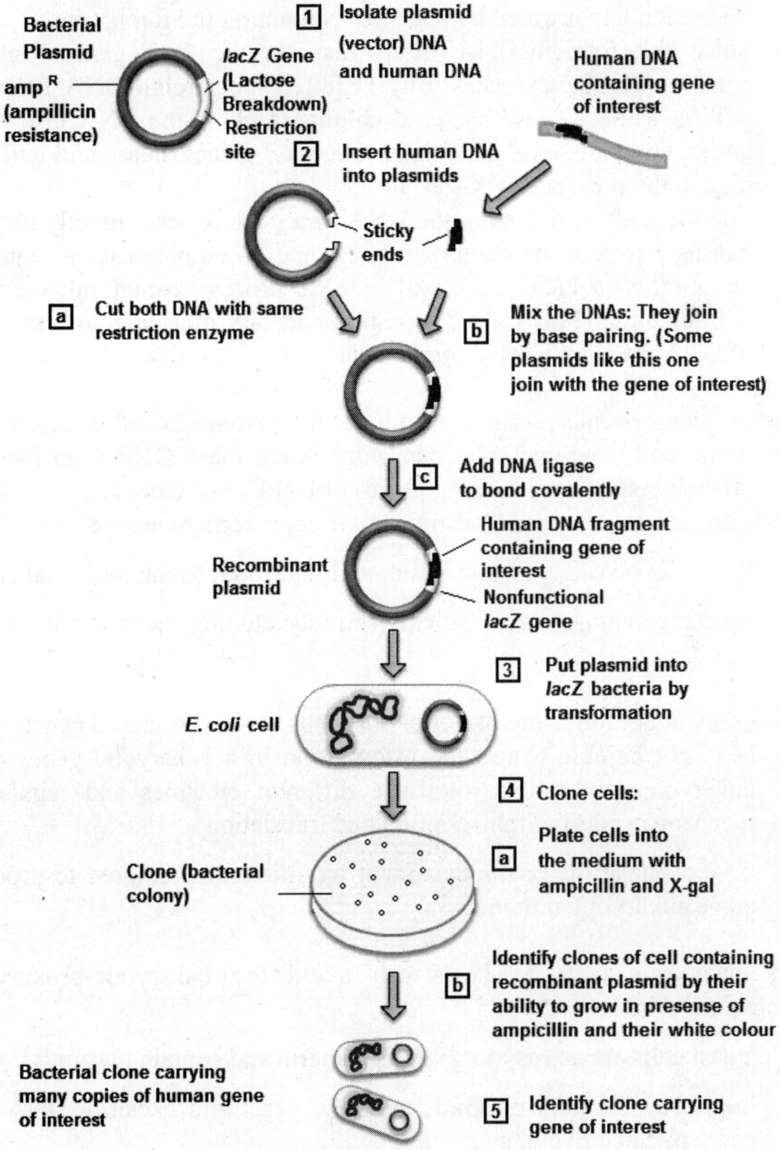

Fig. 1.3: Ligation of vector and insert

***Step 4**: Cloning of cells (and foreign DNA)*

a) Bacteria with the recombinant plasmid are allowed to reproduce, cloning the inserted gene in the process
b) Recombinant plasmids can be identified by the fact that they are ampicillin resistant and will grow in the presence of ampicillin

***Step 5**: Identification of cell clones carrying the gene of interest*

a) X-gal, a modified sugar added to the culture medium, turns blue when hydrolyzed by ß-galactosidase. It is used as an indicator that cells have been transformed by plasmids containing the foreign insert.
b) Since the foreign DNA insert disrupts the *lacZ* gene, bacterial colonies that have successfully acquired the foreign DNA fragment will be white. Those bacterial colonies lacking the DNA insert will have a complete *lacZ* gene that produces ß-galactosidase and will turn blue in the presence of X-gal.
c) The methods of detecting the DNA of a gene depend directly on base pairing between the gene of interest and a complementary sequence on another nucleic acid molecule, a process called nucleic acid hybridization **(Fig. 1.4).** The complementary molecule, a short piece of DNA is called a *nuclei acid probe.*

Artificial chromosomes, which combine the essentials of a eukaryotic chromosome with foreign DNA, can carry much more DNA than plasmid vectors, thereby enabling very long pieces of DNA to be cloned.
Bacteria are commonly used hosts in genetic engineering because:

a) DNA can be easily isolated from and reintroduced into bacterial cells
b) Bacterial cultures grow quickly, rapidly cloning the inserted foreign genes.

Some disadvantages to using bacterial host cells are that bacterial cells:

a) May not be able to use the information in a eukaryotic gene, since eukaryotes and prokaryotes use different enzymes and regulatory mechanisms during transcription and translation
b) Cannot make the posttranslational modifications required to produce some eukaryotic proteins

Using eukaryotic cells as hosts can avoid the eukaryotic-prokaryotic incompatibility issue

a) Yeast cells are as easy to grow as bacteria and contain plasmids.
b) Some recombinant plasmids combine yeast and bacterial DNA and can replicated in either

c) Posttranslational modifications required to produce some eukaryotic proteins can occur

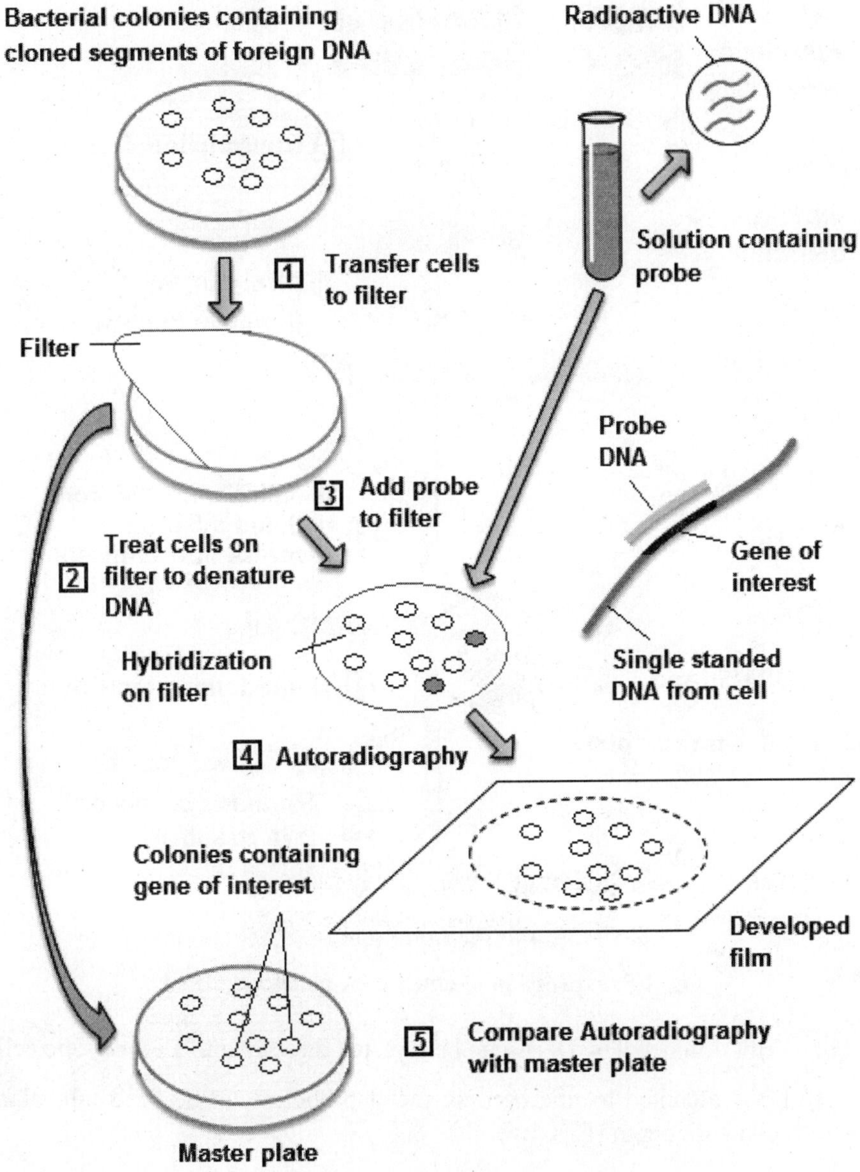

Fig. 1.4: Screening of the clones

Insertion of foreign DNA inside the cells

a) In *electroporation*, a brief electric pulse applied to a cell solution causes temporary holes in the plasma membrane, through which DNA can enter.

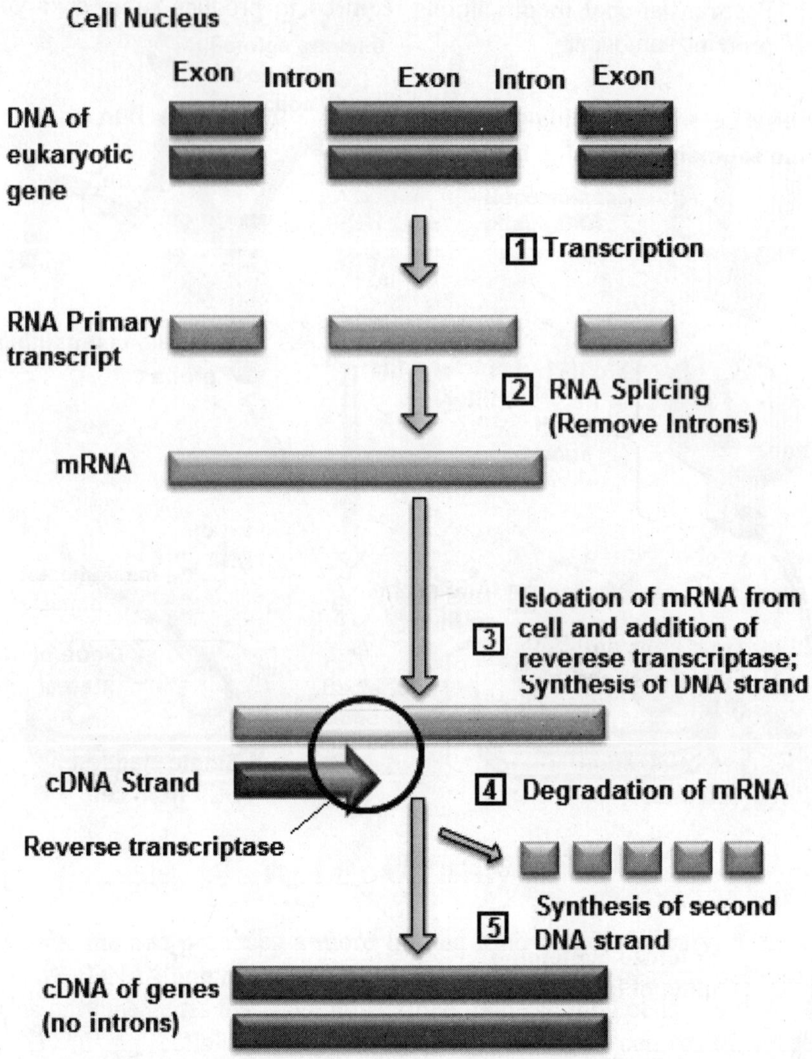

Fig. 1.5: expression of eukaryotic genes via cDNA

b) With thin needles, DNA can be injected directly into a eukaryotic cell

c) DNA attached to microscopic metal particles can be fired into plant cells with a gun **(Fig. 1.6).**

Bacteria and yeast are not suitable for every purpose. For certain applications, plant or animal cell cultures must be used. Cells of more complex eukaryotes carry out certain biochemical processes not found in yeast (e.g. only animal cells produce antibodies)

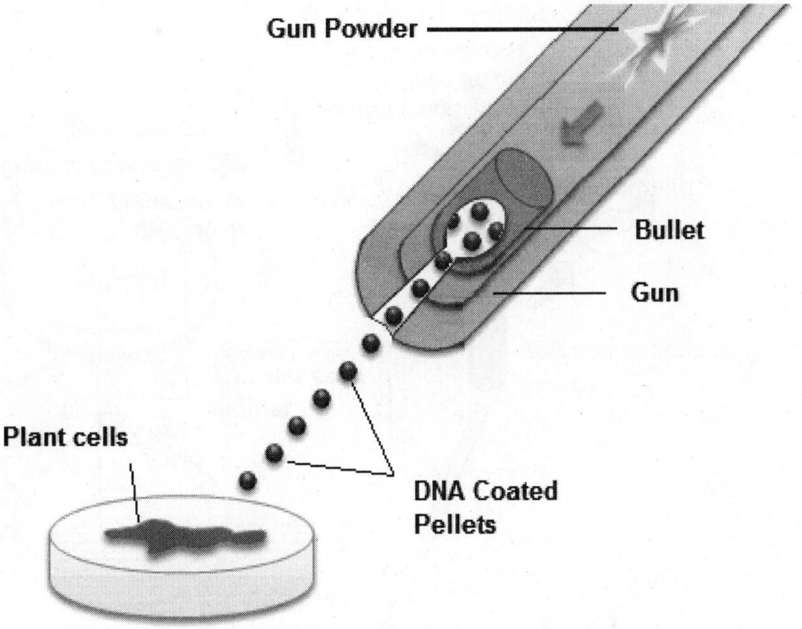

Fig. 1.6: Gene gun

DNA Libraries

Cloned genes are stored in DNA libraries. There are two major sources of DNA which can be inserted into vectors and clones:

a) DNA isolated directly from an organism.

b) Complementary DNA made in the laboratory from mRNA templates.

DNA isolated directly from an organism contains all genes including the gene of interest.

a) Restriction enzymes are used to cut this DNA into thousands of pieces which are slightly larger than a gene.
b) All of these pieces are then inserted into plasmids or viral DNA **(Fig. 1.7)**.
c) These vectors containing the foreign DNA are introduced into bacteria.
d) This produces the *genomic library*, the complete set of thousands of recombinant-plasmid clones, each carrying copies of a particular segment from the initial genome
e) Libraries can be saved and used as a source of other genes of interest or for genome mapping.

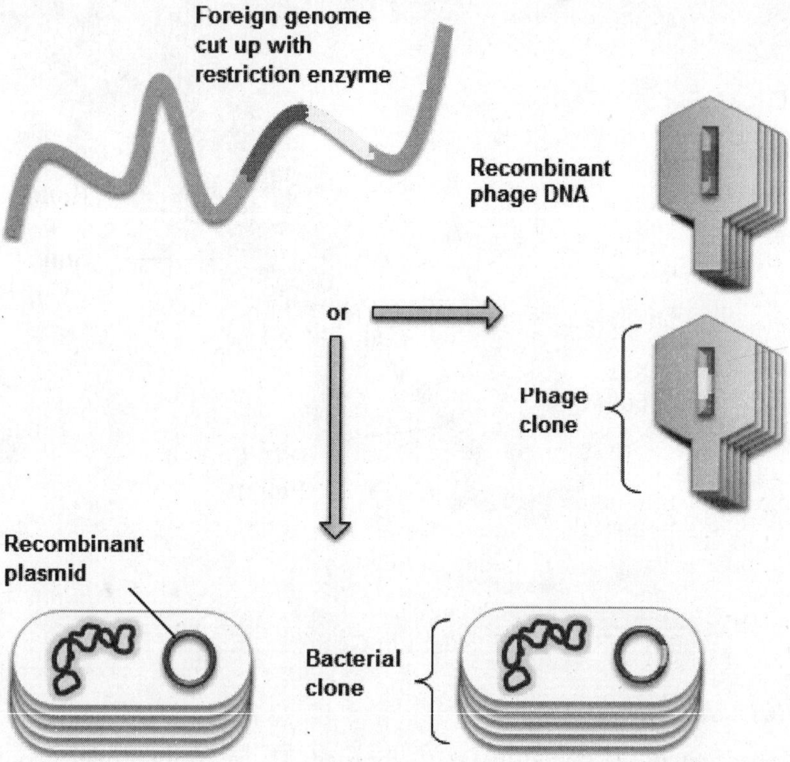

Fig. 1.7: DNA Library

The cDNA method produces a more limited kind of gene library, a cDNA library. A cDNA library represents only part of the cell's genome because it contains only the genes that were transcribed in the starting cells

a) By using cells from specialized tissues or a cell culture used exclusively for making one gene product, the majority of mRNA produced is for the gene of interest
b) For example, most of the mRNA in precursors of mammalian erythrocytes is for the protein hemoglobin.

Polymerase Chain Reaction

The polymerase chain reaction (PCR) clones DNA entirely in vitro. PCR is a technique that allows any piece of DNA to be quickly amplified (copied many times) in vitro conditions **(Fig. 1.8)**.

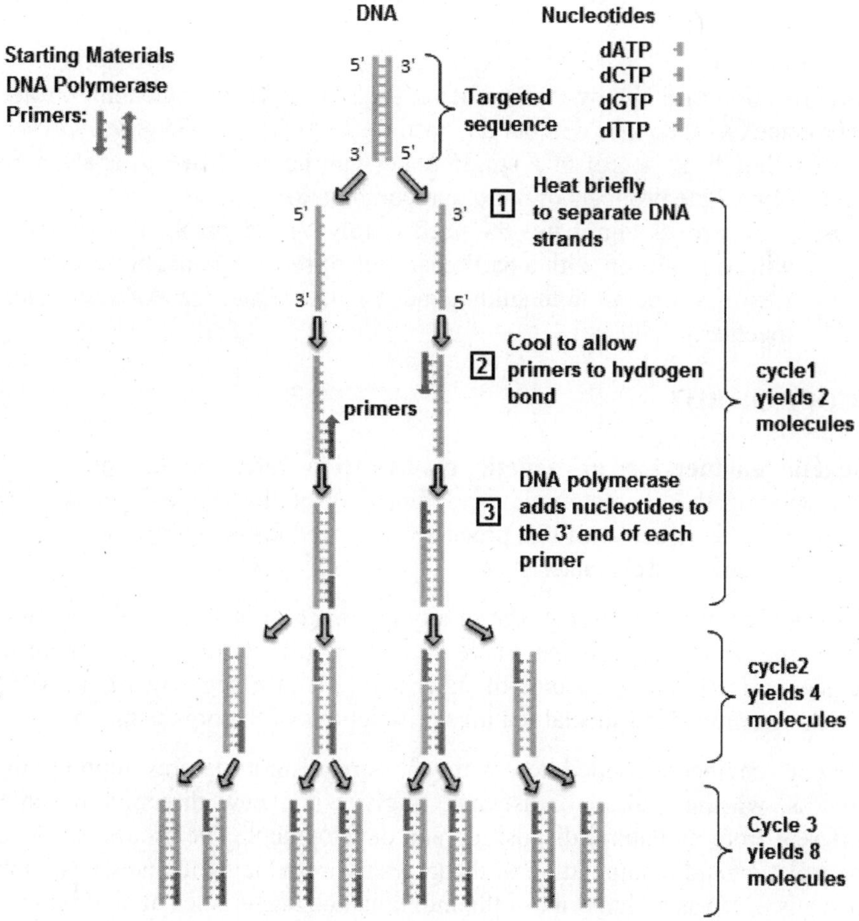

Fig. 1.8: Polymerase chain reaction

a) DNA is incubated under appropriate conditions with special primers and DNA polymerase molecules.
b) Billions of copies of the DNA are produced in just a few hours.
c) PCR is highly specific; primers determine the sequence to be amplified.
d) Only minute amounts of DNA are needed.

PCR is presently being applied in many ways from analysis of DNA from a wide variety of sources:

a) Ancient DNA fragments from a woolly mammoth; DNA is a stable molecule and can be amplified by PCR from sources thousands, even millions, of years old.
b) DNA from tiny amounts of tissue or semen found at crime scenes.
c) DNA from single embryonic cells for prenatal diagnosis.

d) DNA of viral genes from cells infected with difficult to detect viruses such as HIV.

Amplification of DNA by PCR is being used in the Human Genome Project to produce linkage maps without the need for large family pedigree analysis

a) DNA from sperm of a single donor can be amplified to analyze the immediate products of meiotic recombination
b) This process eliminates the need to rely on the chance that offspring will be produced with a particular type of recombinant chromosome.
c) It makes it possible to study genetic markers that are extremely close together.

SCOPE OF RDT

Genetic engineering or genetic modification refers to the process of manipulating the characteristics and functions of the original genes of an organism. The objective of this process is to introduce new physiological and physical features or characteristics.

A **gene** is a basic constituent unit of any organism. It is a locatable region of a genome which contains the whole hereditary information of the organism. A gene corresponds to a unit of inheritance. It is a segment of the **DNA** which determines the special features or functions of the organism.

Genetic engineering meddles with the organism's natural reproductive process, whether sexual or asexual. It gives it a new direction which is different from its natural disposition and development. The process involves the isolation and manipulation of the genes by introducing the new DNA into the cells. DNA is a blue print of the individual characteristics of an organism. The information stored in the DNA controls the management of biochemical process of each organism. The life, development and unique characteristics of the organism depend upon on its own DNA.

The aims of the study, development and practice of **genetic engineering** are noble and beneficial for mankind. Genetic engineering may help make crops resistant to herbicides used to kill the unwanted plants and weeds which obstruct their full growth. Though some herbicides are selective and kill only the specifically targeted unwanted plants, there are others which are non selective and besides killing the useless and obstructive weeds, kill any plants they come in contact thus killing the plants which are sought to be protected.

The domain of genetic engineering can extend from plants to cover both the animal and human life. It can, for example, hybridize the production of the animals and promote the growth of healthy species of milk producing animals, stronger and healthier horses, cows and bullocks which can better withstand the wear and tear of life.

Recreation of vital human organs to replace the sick and missing ones is another example of how genetic engineering can prove beneficial to human beings.

The possibilities of the scope of genetic engineering are limitless and the horizon is widening with every day of research. Genetic engineering like many other branches of science, for example, nuclear science, can be used nobly and ignobly.

The human genome project to categorize all the genes in the human species is a remarkable effort to determine the complete structure of the deoxyribonucleic acid (DNA), the human genetic material, and understand its functions. This research in human genetics aims at determining why the human being resemble or differ from each other.

On the other hand, genetic engineering can also be misused by mutilating the beautiful creation of God by launching, for example, Nazi style schemes for population control or produce a biological devastation through bungling with man made viruses. Cloning is one such example how genetic engineering can be beneficial as well as detrimental.

Molecular biologists have discovered numerous enzymes which can alter the structure of the DNA in living organisms. Using them the scientists can cut the specific genes from the DNA and build customized DNA. With this knowledge they can alter the genes of living organisms.

For example the biological engineers have been able to change the growth patterns of tomatoes. Tomatoes are sensitive to frost that shortens their life span. Fish on the other hand thrive in the cold and chilly water. Scientists found out the gene in fish that resists the cold and transferred it to tomatoes thereby immunizing them against the frosty cold and prolonging their growing season. There are innumerable such projects that are creating new strains in the agricultural areas which have great economic potential.

Genetic engineering can boost the growth and yield of a variety food crops–pulses and cereals–and alleviate the problems of food shortage and mitigate the suffering of the starving millions across the world.

Genetic engineering is not just an extension of conventional breeding. In fact, it differs profoundly. As a general rule, conventional breeding develops new plant varieties by the process of selection, and seeks to achieve expression of genetic material which is already present within a species. (There are exceptions, which include species hybridization, wide crosses and horizontal gene transfer, but they are limited, and do not change the overall conclusion, as discussed later.) Conventional breeding employs processes that occur in nature, such as sexual and asexual reproduction. The product of conventional breeding emphasizes certain characteristics. However these characteristics are not new for the species. The characteristics have been present for millennia within the genetic potential of the species.

Genetic engineering works primarily through insertion of genetic material, although gene insertion must also be followed up by selection. This insertion process does not occur in nature. A gene "gun", a bacterial "truck" or a chemical or electrical treatment inserts the genetic material into the host plant cell and then, with the help of genetic elements in the construct, this genetic material inserts itself into the chromosomes of the host plant. Engineers must also insert a "promoter" gene from a virus as part of the package, to make the inserted gene express itself. This process alone, involving a gene gun or a comparable technique, and a promoter, is profoundly different from conventional breeding, even if the primary goal is only to insert genetic material from the same species.

But beyond that, the technique permits genetic material to be inserted from unprecedented sources. It is now possible to insert genetic material from species, families and even kingdoms which could not previously be sources of genetic material for a particular species, and even to insert custom-designed genes that do not exist in nature. As a result we can create what can be regarded as synthetic life forms, something which could not be done by conventional breeding.

It is interesting to compare this advance to the advances that led to creation of synthetic organic chemicals earlier in the 1900s. One could argue that synthetic chemicals are just an extension of basic chemistry, and in certain senses they are. Yet when we began creating new chemicals that had not previously existed on the earth, or which had only been present in small quantities, and began distributing them massively, we discovered that many of these chemicals, even though they were made of the same elements as "natural" chemicals, had unexpected adverse properties for the environment and health. Because we had not co-evolved with them for □millennia, many (though by no means all) had negative effects. Among the serious problems were PCBs and vinyl chloride, which were found to be carcinogens, and numerous organochlorine pesticides, which were found to be carcinogens, reproductive toxins, endocrine disruptors, immune suppressors, etc. After several decades of use, these effects caused such concern that we passed the Toxic Substances Control Act which required premarket screening of synthetic organic chemicals by EPA for such effects as carcinogenicity, mutagenicity and impact on wildlife, and changed our pesticide rules similarly. There are many ways in which these two scientific advances are not analogous, but the experience with synthetic organic chemicals underlines the potential for unexpected results when novel substances are introduced into the biosphere.

GENETIC ENGINEERING GUIDELINES

With the success of Boyer-Cohen experiments (in 1973), it was realized that recombinant DNA technology could be used to create organisms with novel genes. This created worldwide commotion (among scientists, public and government officials) about the safety, ethics and unforeseen consequences of genetic manipulations. Some of the phrases quoted in the media during those days are given:

1. Manipulation of life
2. Playing God
3. Manmade evolution
4. The most threatening scientific research

It was feared that some new organisms created inadvertently or deliberately for warfare, would cause epidemics and environmental catastrophes. Due to the fears of the dangerous consequences, a cautions approach on recombinant DNA experiments was suggested.

In 1974, a group of ten scientists led by Paul Berg wrote a letter that simultaneously appeared in the prestigious journals Nature, Science and Proceedings of the National Academy of Sciences. The dangers of DNA technology were printed out in that letter (highlights given below):

"Recent advances in techniques for isolation and rejoining of segments of DNA now permit construction of biologically active recombinant DNA molecules ***in vitro***. Although such experiments are likely to facilitate the solution of important theoretical and practical biological problems, they would also result in creation of novel types of DNA elements whose biological properties cannot be completely predicted. There is a serious concern that some of these DNA molecules could prove biologically hazardous". The letter also appealed to molecular biologists worldwide for a moratorium on many kinds of recombinant DNA research, particularly those involving pathogenic organisms.

Asilomar Recommendations

In February 1975, a group of 139 scientists from 17 countries held a conference at Asilomar, a conference centre in California, USA. They assured the uneasy public that the microorganism used in DNA experiments were specifically bred and could no survive outside the laboratory. These scientists formulate guidelines and recommendations for conducting experiments in genetic engineering

NIH Guidelines

National Institute of Health (NIH), USA, constituted the Recombinant DNA Advisory Committee (RAC) which issued a set of stringent guidelines to conduct research on DNA. RAC was in fact overseeing the research projects involving gene splicing and recombinant DNA.

Some of the important original NIH recommendations on recombinant DNA research relate to the following aspects.

(i) Physical (laboratory) containment levels for conducting experiments.

(ii) Biological containment–the host into which foreign DNA is inserted should not proliferate outside the laboratory or transfer its DNA into other organisms.

(iii) For research on pathogenic organisms, elaborate, controlled and self-contained rooms were recommended.

(iv) For research on less dangerous organisms, units equipped with high quality filter systems should be used.

(v) No deliberate release of any organism containing recombinant DNA into the environment.

It may be noted here that although the NIH guidelines did not have legal status, most institutions, companies and scientists voluntarily complied.

Relaxation of NIH Guidelines

It was in the 1980s that the original NIH guidelines were considerably relaxed by NIH-RAC, based on the experience and experimental data obtained from the NIH-sponsored studies on recombinant DNA research. It is a fact that the genetic engineering research flourished and progressed rapidly after relaxation of NIH guidelines. It may however be noted that NIH-RAC continues to be a watchdog over the DNA technology experiments

PHARMACEUTICAL PRODUCTS OF RECOMBINANT DNA

As the recombinant DNA technology progressed, many pharmaceutical compounds of human health care are being produced though genetic manipulations. Most countries consider that the existing regulations for approval of pharmaceuticals of commercial use are adequate to ensure safety since the process by which the product is manufactured is irrelevant. Thus, the recombinant DNA product (protein, vaccine, and drug) is evaluated for its safety and efficacy like any other pharmaceutical product.

GENETICALLY ENGINEERED ORGANISMS (GEOS)

Recombinant DNA research has resulted in the creation many genetically engineered organisms. These include microorganisms, animals and plants. The latter two respectively result in transgenic animals and transgenic plants.

THE FUTURE OF GENETIC ENGINEERING

DNA technology has largely helped scientists to understand the structure, function and regulation of genes. The development of new/modern biotechnology is primarily, based on the success of DNA technology. Thus, the present biotechnology (more appropriately molecular biotechnology) has its main roots in molecular biology. Biotechnology is an interdisciplinary approach for applications to human health, agriculture, industry and environment. The major objective of biotechnology is to solve problems associated with human health, food production, energy production and environmental control.

It is an accepted fact that recombinant DNA technology has entered the mainstream of human life and has become one of the most significant applications of scientific research. Biotechnology is regarded as more an art than a science. After the successful sequencing of human genome, many breakthroughs in biotechnology are expected in future.

PATENTING OF BIOLOGICAL MATERIAL

If we discuss the different intellectual properties and the rights to protect these properties, particularly when they involve biotechnological processes or products, of these properties, patents are the most important. Even though, patenting of inventions arising from basic research (commercially significant) was earlier considered unfashionable or ethically dubious, it is now viewed more favorably by many. Different countries have different patent laws, which can be modified from time to time.

Patents are granted or complaints of infringement of these patents decided by courts in accordance with the patent law of the concerned country. Before 1980, when the discovery of an oil-eating bacterium (*Pseudomonas*) by a nonresident Indian scientist (Dr. Chakrabarty), was patented in USA by a multinational corporation, the life forms could not be patented. A later patent issued for 'oncomouse' was another milestone in patenting of life forms.

Recently the issue of granting patents to life forms has also been discussed in India. Even though, there are arguments against patenting life forms particularly in developing world, but the advent of biotechnology has made it necessary that patents of life forms be allowed, because in its absence, commercial firms would not like to invest in biotechnology research. In this connection, there has been a lot of debate in India, because in recent years, USA has been pressurizing India through the multinational forum of GATT (General Agreement of Trade and Tariffs) to change many of its laws pertaining to patents.

In view of recent biotechnological developments, the discussion has now progressed beyond merely the question of-whether living organisms can be patented, and is now concerned with establishing solid criteria for granting patents. The existing patent laws are also being reinterpreted with regard to biological material.

International Conventions on Biological Material Patenting

In connection with intellectual properties dealing with biological material, there are international conventions, which do not allow patents for processes or products dealing with alleviation of human diseases.

For instance, it is not possible to patent techniques for surgery such as by pass heart surgery, organ transplants, implant of heart valves, artificial limbs, dialysis for renal failure, plastic surgery and other surgical methods used in cancer, appendicitis or neurosurgery.

Similarly, one cannot patent the following:

(i) The use of drugs, antibiotics or vaccines for any form of diagnosis, prevention or cure of diseases.
(ii) Artificial insemination.
(iii) In vitro fertilization and embryo transfer, etc.

In plant biotechnology also, live plants (not the transgenic plants), naturally occurring microorganisms, micro propagation, tissue and organ culture techniques, biological control of pests or hybrid varieties can not be protected using patents. Even techniques for cell fusion, protoplast fusion and gene transfer can not be generally protected through patents.

Commercial benefits from hybrid varieties can, however, be secured through control of parent plants, which may be a trade secret. Similarly, although new crop varieties can not be patented, but transgenic animals and plants can be patented, although there may be arguments against it.

In case of microorganisms selected for production of antibiotics, amino acids, enzymes, alcohol, etc., patent protection is generally not available, although under specific situations, protection for microorganisms modified for commercial use may be allowed.

In these cases (e.g. fermentation), generally commercial benefits are derived by contracts involving transfer of technology through restrictions on further transfer to third parties. Thus, trade secrecy rather than patents in such cases protect the commercial benefit.

International Co operations on Biological Material Patenting

International cooperation has also been sought for application of patent law to solve problems, which are not confined to one or few countries. In a convention called Paris Convention of 1988, the basic principles of equal treatment for domestic and foreign inventors were established.

The Paris Convention now has 100 member states. The convention allows inventors to claim international priority by filing of a patent application initially in one member state and subsequently in others.

The main instrument of international collaboration for intellectual property is the World Intellectual Property Organization (WIPO) based in Geneva. It administers (but not enforces law) the Paris Convention and all subsequent conventions. WIPO operates by asking member states to ratify a convention and to introduce the agreed basic principles into their national laws.

Another international group is the European Patent Convention (EPC) of 1973, which began operation in 1978 and has 14 member states. EPC has the distinction of being the first patent statute to introduce specific provisions for biotechnology inventions. Two of these important provisions include the following:

(i) Need of culture collections as patent depositories for the placement of microorganisms referred to in patent applications; this is necessary, because the complete description of living material is difficult and may not be fully reproducible.

(ii) Exclusion of certain plant related inventions from the list of those which can be patented, e.g. plant and animal varieties bred through classical methods.

First major study of international patent protection for biotechnology was published in 1985 by the Organization for Economic Cooperation and Development (OECD). WIPO also began a wider study through experts on biotechnological inventions, with an objective to provide suggested solutions for problems related to patent law.

Taking EPC as modes in 1988, the European Commission (EC) formulated a European Commission Directive, to help the member states modify their national laws, so that they become uniform in European Countries. This directive has following three major proposals, which will make patent laws more favorable for patent applicants:

(i) An invention must not be refused protection simply because it involves living matter (earlier no patents were granted for living forms). Examples of patentability of microbiological processes and products.

(ii) The Scope of patent should extend to all progeny produced by multiplication of parental material, provided it retains the characteristics of this patented material. Generally, under patent law, there is provision of exhaustion of rights once the product is sold, so that the purchaser can cultivate unlimited quantity from a small quantity purchased from the patent holder (this is overcome in the directive).

Further, the patented invention can be used for experimental use without infringement of the patent. But the experiments may sometimes be designed as a way around the patent and may thus be utilized for commercial purposes. Thus if this directive is accepted, patented living forms will neither be allowed multiplication by the purchaser, nor will it be allowed to be used for experiments without paying the royalty for the patent.

(iii) The Microorganisms to be patented will have to be deposited in culture collections. For convenience, only microorganisms are used for illustration, but the categories shown in this also apply equally to other types of biological material (e.g. cell lines, plant and animal cells).

Patents for Higher Plants and Higher Animals

Under patent laws of most countries, a patentable invention must be capable of industrial application. Agriculture is also treated as an industry in this connection, so that a wide range of agricultural and horticultural methods and products can be patented, provided these arc inventions and represent novelties.

Biopesticides and Bioinsecticides arc such examples. Extensive patent literature is also available on the use of *Bacillus thuringiensis* and mycoherbicides (fungi). Other examples include:

(i) Novel techniques of plant micro propagation, and
(ii) Plant cell and tissue culture methods to prepare useful metabolites.

Among higher plants, an important example is the patent for'tryptophan overproducing maize' obtained through tissue culture. The patent known as 'Hibberd Patent' was issued to 'Molecular Genetics Research and Development'.

A similar patent was allowed in US for a transgenic animal, popularly described as 'oncomouse patent', Patent was also allowed for a polyploid oyster produced by the application of hydrostatic pressure to zygotes.

In view of these cases, Patent and Trademark Office now considers claims for patents of 'non naturally occurring non human multicellular living organisms'. Japanese patent law is not very different from the US law and patents for animals and plants have been granted in Japan also, provided these were products of specific methods or inventions.

In Europe also patents are allowed for such plants and animals, even though plant varieties and animal varieties arc excluded. In this connection there have also been disputes regarding what constitutes a variety. Patent claim for a soybean variety applied by Pioneer Hi-bred, was rejected by Supreme Court of Canada, on the ground that no description of the method was available, although the seeds of the variety were deposited.

In some cases hybrid varieties were also considered patentable. It is not certain whether the meaning of a 'variety' will be restricted, or if the scope of a variety will be wider. This is an area which is receiving major attention in several countries and will be resolved in the coming years.

Patenting Transgenic Organisms and Isolated Genes

Transgenic plants and animals can be protected through patent claims in several countries now including USA, Japan, Europe, etc. 'Oncomouse' is one such example, where product claim was initially rejected by EPO, but the decision has been overruled on appeal. This is despite the fact that animal genetic manipulation may be considered by some to be an ethical question.

Among transgenic plants, herbicide resistant cotton, insect resistant tobacco and virus resistant potato have been patented. Boll worm resistant cotton may also be allowed patent protection (consult Sci. Amer., March, 1991). There is no doubt, therefore, that 10 future patent claims for transgenic plants and animals fulfilling certain conditions will have to be allowed in several countries.

Patenting of Genes and DNA Sequencing

The genes that are synthesized artificially, normally fulfill the requirements of patents and, therefore, have no difficulty in gelling patent protection. If the protein, that the artificial gene makes, and the organism into which the gene is inserted arc also novel having desirable attributes such a patent may extend to the protein and the organisms also.

Thus the patents involving artificial genes and novel higher organisms (plants or animals) are relatively straight forward. In recent years, one of the most widely discussed political issue in biotechnology concerns the patenting of naturally occurring useful genes, because they do not fulfill the requirement of novelty (they are only discovered, not invented and for a patent, invention rather than discovery is necessary)

The developing countries have been arguing that patenting of naturally occurring genes will allow the developed countries to make use of a weed from their land, isolate a desirable gene from this weed, transfer it to a commercial crop and then sell it back to the same country (which was source of the gene).

There are counter arguments, that companies should be rewarded for their investment and labor involved in making the natural genes useful, which may otherwise ever be exploited for human welfare. The law on the above issue is both confusing and changing through judgments on disputes by courts.

However, the law in U.S.A allows patent on purified form of a chemical, if in nature it occurs in an impure form. In UK, on the other hand, the courts recently held that a naturally occurring gene sequence can not be patented. It is, however, now certain that in U.S.A and some other developed countries, patents will be allowed in future for genes isolated and cloned from nature.

An example of such a patent granted involved cloning of d DNA fragment, which originated due to mutation in a microorganism and imparted resistance to the herbicide glyphosate. Intended for integration into plants to confer glyphosate resistance, the gene is patented by Calgene Inc. in USA in terms of a DNA sequence containing the relevant structural gene. Increasing number of such patents arc now appearing particularly in USA, and similar applications for patents arc awaiting decision in Europe and Japan. Therefore, whenever a patent for a gene is allowed, biotechnology companies would like the protection to extend to the marketed products. This area of IPR in biotechnology will receive further increased attention in coming years.

In 1992, the question of patenting isolated genes or DNA and cDNA sequences has once again been addressed in USA, due to an application on behalf of National Institute of Health (NIH), USA, filed for patenting hundreds of cDNA sequences. The main issue for discussion was whether these sequences are useful as required in patent statute.

This requirement, in 1966 (Brenner vs. Manson, US Supreme Court, 148 USPQ 689, 1966) was interpreted to mean that in the process or product to be patentable, there should exist a specific benefit in currently available form. Subsequently the word useful was interpreted to mean practical utility. Many of the isolated genes may not offer specific benefit in its currently available form and may not be of practical utility.

This will render those unpatentable, even though it may discourage research in this area. Despite the above requirement, isolated genes, vectors and transformed cells expressing the hormone – angiogenesis factor (AGF), which increases vesicularization, has been allowed a patent. This patent was allowed, even though at the time of application (1985), AGF did not really have practical utility for therapy or for diagnostic tests. Many other patents of isolated genes are similar in nature and their validity may be questionable if challenged in a law suit. Such law suits, challenging the validity of parents for isolated genes already granted, will settle this issue in future. Patenting in such cases may be necessary to encourage research leading to isolation of genes, which if patented, will be available to other researchers and if not patented will be kept a secret.

Therefore, patent applications for isolated genes may in future; either face a rigid attitude asking for specific practical utility in its currently available form or may receive a relaxation in view of its utility in immediate research with future potential of practical utility

Plant Breeders' Rights

Plant varieties are generally protected in several countries (not in India) through plant breeders right (PBR) or plant variety rights (PYR). Through these rights, further propagation of the variety is restricted.

Under the existing convention due to International Union for the Protection of New Varieties (UPOY), the breeder's right does not prohibit the farmer from reuse (plant back) of farm saved seed of a variety from his own harvest for planting another crop.

Furthermore, the protected plant variety can be freely used as a plant genetic resource for the purpose of breeding other varieties. Since the revised UPOV convention extends PBR to cover the reuse, countries arc now free to extend PBR, to cover the reuse of seed, so that farmers in such a case, can not use his own seed without paying a royalty to the PBR holder.

However, most countries are expected to limit the PBR with regard to farmer's plant back, although a farmer can not sell the seed. Enforcement of such rights as above, in case of freely reproducible material, is only possible with large holdings of land, as in case of plantations or in very large farms, and in case of high value cash crops (e.g. cashew, species, medicinal plants, etc.).

Even in these cases, infringements are difficult to prove. When patent or plant breeders rights arc not available for true breeding crop varieties, plant breeder (particularly private plant breeder in developed countries like Germany) may feel tempted to focus their efforts on developing hybrid varieties.

Because hybrids do not breed true and give higher yields, no one would raise a crop from harvested seed that will give reduced yield. Thus hybrid varieties may give to the plant breeder an advantage, which is equivalent to intellectual property protection.

PBR has analogies to patents, but there arc also important differences. Rights are granted for a limited period (usually 20 years) to the breeder. Breeder seeking PBR can not seek exclusive rights for a unique feature of his variety, although under patent this is allowed. For instance, a breeder of the first blue rose can not monopolize blue color of rose.

It will be open to other breeders to breed and protect blue roses, which are distinct from the first variety having blue roses. A protected variety must fulfill some requirements. It should be:

a. New.
b. Distinct.
c. Uniform.
d. Stable.

The third option allows the present practice to continue, where complete freedom for introduction and promotion of new varieties is allowed. This option seems to be the best for Indian Agriculture all ease for the time being although there is considerable pressure from developed countries to accept some form of IPR. It is possible, that India may have to introduce some sort of PBR for crop varieties. This will encourage private companies to enter plant breeding programmes.

New means, the variety should not have been previously exploited commercially. Distinct means, it should he dearly distinguishable from all other varieties known at the date of application for protection. Uniform means that all plants of the variety should he sufficiently uniform. Stable means that the variety can be reproduced and multiplied without losing its characteristics and uniformity.

In India, new crop varieties are bred at state Agriculture Universities and at state Departments of Agriculture. The seed of new crop varieties flows freely to farmers and to the private compal1ies and no royalty is payable. This really encouraged farmers, in the past, to grow new varieties leading to green revolution.

2

Enzymes in Genetic Engineering

Some technological developments in science enable us to gain immense knowledge and increased potential for innovation. Genetic engineering and biomedical research have experienced such a revolutionary change since the past 30 years with the development of gene manipulation. The ability to manipulate DNA in vitro (outside the cell) depends entirely on the availability of purified enzymes that can cleave, modify and join the DNA molecule in specific ways. At present, no purely chemical method can achieve the ability to manipulate the DNA in vitro in a predictable way. Only enzymes are able to carry out the function of manipulating the DNA. Each enzyme has a vital role to play in the process of genetic engineering.

HISTORY

Today, in the age of molecular biology, the study of an organism's genome (its complete DNA) is a central component driving our understanding of biology. When scientists first considered studying genomes they were faced with a problem: how to reproducibly cut a genome's DNA into fragments that were small enough to handle? It was a significant problem. Genomes are composed of large DNA chunks on the order of millions of units, while a scientist could reasonable only handle pieces of DNA a few thousand units long. A discrepancy far too large to bridge, thus a method for reproducibly cutting DNA into manageable pieces was required to move the genomic studies forward. When this question was first posed in the 1970s it was becoming a relatively simple exercise to isolate DNA and then randomly cut it up using chemical or mechanical means. Unfortunately, this random cutting was not a satisfactory way to obtain smaller pieces of DNA, since it was impossible to tell what the original order of the DNA fragments were, an important point since the specific order of DNA is essential for its function. The biologists were stuck. A breakthrough was needed.

As with many of the important discoveries in biology, it was the study of bacteria that yielded this breakthrough. It was discovered that a type of bacterial enzyme was found to have the ability to cut DNA in a test tube. These restriction endonuclease, so named because they cut double stranded

DNA at restricted sites, were discovered as a natural part of the bacterial machinery. In a bacterial cell, restriction endonucleases (often referred to as restriction enzymes) act as a kind of immune system, protecting the cell from the invasion of foreign DNA, as would occur when a virus attempted to infect a bacterial cell. These restriction endonucleases provided biologists with a tool to study and manipulate DNA by enabling the generation of consistently sized DNA fragments. They are now used for a wide range of applications, including cloning, Southern hybridization analysis, DNA sequencing and global gene expression analysis (SAGE). Truth be told, many recombinant DNA technologies, which the field of biotechnology heavily relies on, are unlikely to have been developed without the discovery of restriction enzymes.

NUCLEASES

Nucleases are a group of enzymes which cleave or cut the genetic material (DNA or RNA).

DNase and RNases

Nucleases are further classified into two types based upon the substrate on which they act. Nucleases which act on or cut the DNA are classified as DNases, whereas those which act on the RNA are called as RNases.

DNases are further classified into two types based upon the position where they act. DNase that act on the ends or terminal regions of DNA are called as Exonucleases and those that act at a non-specific region in the centre of the DNA are called as Endonucleases. Exonucleases require a DNA strand with at least two 5 and 3 ends.

They cannot act on DNA which is circular. Endonucleases can act on circular DNA and do not require any free DNA ends (i.e. 5 or 3 end). Exonucleases release nucleotides (Nucleic acid + sugar + phosphate), whereas endonucleases release short segments of DNA.

Restriction Enzymes

DNases which act on specific positions or sequences on the DNA are called as restriction endonucleases. The sequences which are recognized by the restriction endonucleases or restriction enzymes (RE) are called as recognition sequences or restriction sites. The recognition sequences for the vast majority of type II restriction endonucleases are normally palindromes that is the sequence of bases is the same on both strands when read in the 5′ → 3′ direction, as a result of a twofold axis of symmetry. In some cases, the cleavage points occur exactly on the axis of symmetry, giving products (restriction fragments) which are **blunt-ended**. In most cases, however, the cleavage points do not fall on the symmetry axis, so that the resulting restriction fragments possess so-called 5′ overhangs or 3′ overhangs **(Fig. 2.1).**

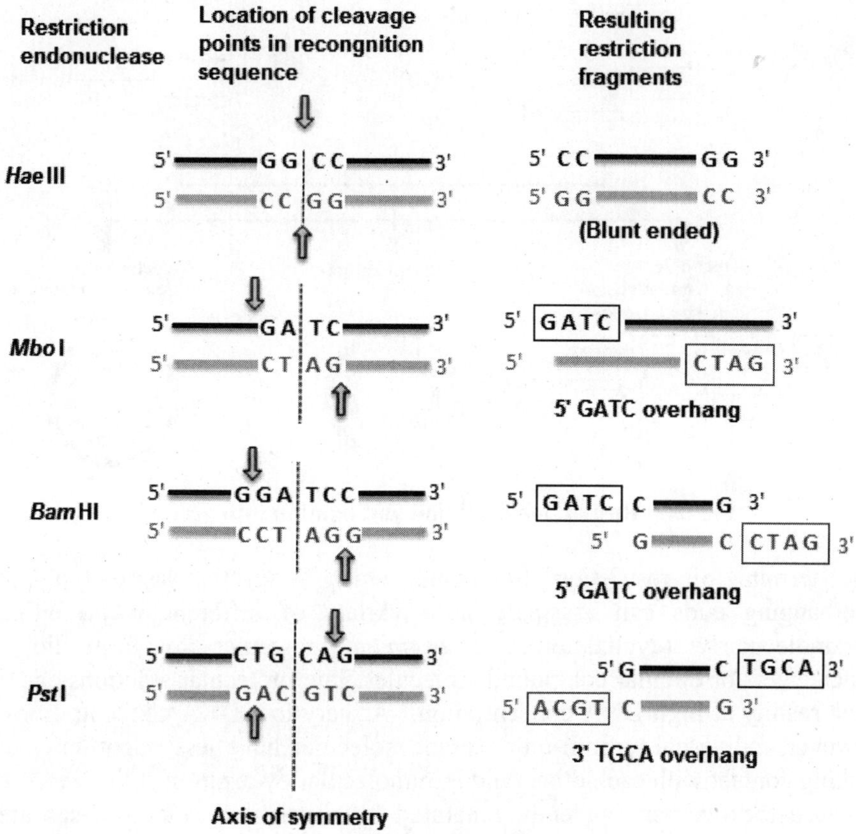

Fig. 2.1: Restriction Digestion

Overhanging ends generated by cleavage with a restriction nuclease are often described as sticky ends or **cohesive termini** because the two overhanging ends of each fragment are complementary in base sequence, and will have a tendency to associate with each other, or with any other similarly complementary overhang, by forming base pairs. Different fragments with the same sequences in their overhanging ends can be generated by: (i) cutting with the same restriction nuclease; (ii) cutting with different restriction endonucleases that happen to recognize the same target sequence (**isoschizomers**); or (iii) by cutting with enzymes which have different recognition sequences but happen to produce compatible sticky ends, for example *Bam*HI and *Mbo*I **(Fig. 2.2)**.

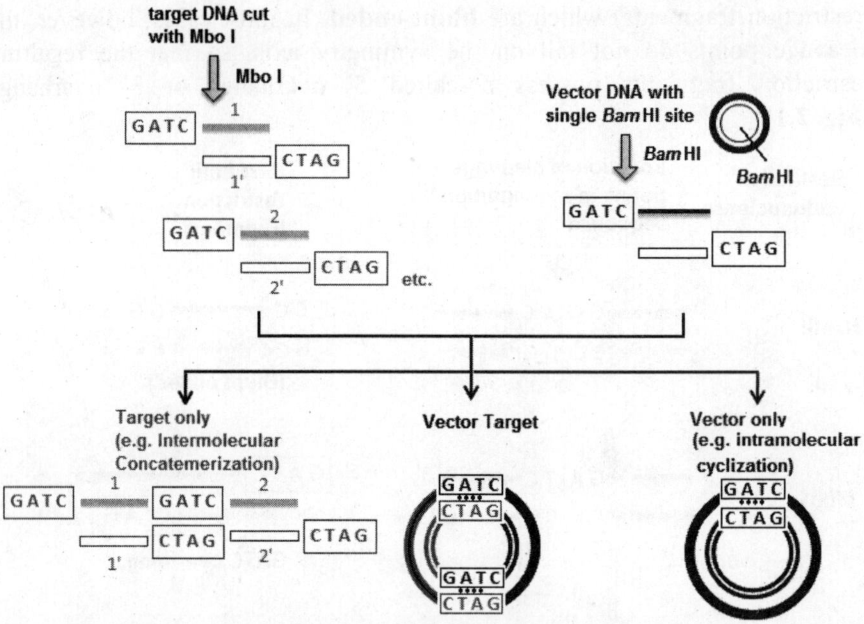

Fig. 2.2: Restriction digestion and ligation into vector

The termini of restriction fragments which have the same type of overhanging ends can associate in a variety of different ways, either intramolecularly (cyclization), or between molecules to form linear concatemers or circular compound molecules. Intermolecular reactions occur most readily at high DNA concentrations. At very low DNA concentrations, however, individual termini on different molecules have less opportunity of making contact with each other, and intramolecular cyclization is favored. Because the overhanging ends generated by restriction endonucleases are very short (typically four nucleotides or less), hydrogen bonding between complementary overhanging ends provides a rather weak contact between two molecules, and can only be maintained at low temperatures. However, it does facilitate subsequent covalent bonding between the two associated molecules (DNA ligation). This is performed using the enzyme DNA ligase. Ligation of blunt-ended fragments is also possible, although less efficient than sticky end ligation.

Generally, ligation reactions are designed to promote the formation of recombinant DNA (by ligating target DNA to vector DNA), although vector cyclization, vector-vector concatemers and target DNA-target DNA ligation are also possible. To achieve this, the vector molecules are often treated so as to prevent or minimize their ability to undergo cyclization. There are two common ways of achieving this.

Cutting of vectors with two different restriction endonucleases

Often, vector molecules have multiple unique restriction sites, in which foreign DNA can be cloned, occurring in a short segment of the molecule. It is often convenient, therefore, to cut the vector with two restriction endonucleases which do not produce complementary overhanging ends (e.g. *Eco*RI and *Bam*HI), and remove the small vector fragment between the sites, resulting in a vector molecule whose two ends cannot religate. However, if target DNA is cut with the same enzyme combination, recombinant DNA can easily be formed.

Vector dephosphorylation

During DNA ligation *in vitro*, the enzyme DNA ligase will catalyze the formation of a phosphodiester bond only if one nucleotide contains a 5′ phosphate group and the other contains a 3′ hydroxyl group. If the 5′ phosphate groups at both ends of the vector DNA are removed by treatment with alkaline phosphatase, the tendency for the vector DNA to recircularize will therefore be minimized. A foreign DNA insert can, however, provide 5′-terminal phosphates which can then be joined to the 3′ hydroxyl groups provided by the vector. This method, therefore, increases the frequency of cells containing recombinant DNA.

Types of Restriction endonucleases

There are three types of restriction enzymes: Type I, Type II and Type III.

Type I Restriction Enzymes

These restriction enzymes recognize the recognition site, but cleave the DNA somewhere between 400 base pairs (bp) to 10,000 bp or 10 kbp right or left. The cleavage site is not specific. These enzymes are made up of three peptides with multiple functions. These enzymes require Mg++, A TP and S adenosyl methionine for cleavage or for enzymatic hydrolysis of DNA. These enzymes are studied for general interest rather than as useful tools for genetic engineering.

Type II Restriction Enzymes

Restriction enzymes of this type recognize the restriction site and cleave the DNA within the recognition site or sequence. These enzymes require Mg++ as cofactor for cleavage activity and can generate 5 -PO_4 or 3 -OH. Enzymes of this type are highly important because of their specificity. Type II restriction enzymes are further divided into two types based upon their mode of cutting:

1. Blunt End Cutters

Blunt end cutters Type II restriction enzymes cut the DNA strands at same points on both the strands of DNA within the recognition sequence. The

DNA strands generated are completely base paired. Such fragments are called as blunt ended or flush ended fragments.

2. *Cohesive End Cutters*

Cohesive end cutter Type II restriction enzymes of this class cut the DNA stands at different points on both the strands of DNA within the recognition sequence. They generate a short single-stranded sequence at the end. This short single strand sequence is called as sticky or cohesive end. This cohesive end may contain 5 -PO_4 or 3 -OH, based upon the terminal molecule (5 -PO_4 or 3 -OH).These enzymes are further classified as 5end cutter (if 5 -PO_4 is present) or 3 -end cutter (if 3' -OH is present).

Type III Restriction Enzymes

Type III Restriction enzymes of this type recognize the recognition site, but cut the DNA 1 kbp away from the restriction site. These enzymes are made up of two peptides or subunits. These enzymes require A TP, Mg++ and S-adenosyl methionine for action.

R-M Systems

Restriction enzymes usually occur in combination with one or two modification enzymes (DNA-methyltransferases) that protect the cell's own DNA from cleavage by the restriction enzyme. Modification enzymes recognize the same DNA sequence as the restriction enzyme that they accompany, but instead of cleaving the sequence, they methylate one of the bases in each of the DNA strands. The methyl groups protrude into the major groove of DNA at the binding site and prevent the restriction enzyme from acting upon it. Together, a restriction enzyme and its "cognate" modification enzyme(s) form a restriction-modification (R-M) system. In some R-M systems the restriction enzyme and the modification enzyme(s) are separate proteins that act independently of each other. In other systems, the two activities occur as separate subunits, or as separate domains, of a larger, combined, restriction-and-modification enzyme.

Mode of Action of Restriction Enzymes

The restriction enzyme binds to the recognition site and checks for the methylation (presence of methyl group on the DNA at a specific nucleotide). If there is methylation in the recognition sequence, then, it just falls off the DNA and does not cut. If only one strand in the DNA molecule is methylated in the recognition sequence and the other strand is not methylated, then RE (only type I and type III) will methylate the other strand at the required position. The methyl group is taken by the RE from S-adenosyl methionine by using modification site present in the restriction enzymes.

However, type II restriction enzymes take the help of another enzyme called methylase, and methylate the DNA. Then RE clears the DNA. If there is no methylation on both the strands of DNA, then RE cleaves the DNA. It is only by this methylation mechanism that, RE, although present in bacteria, does not cleave the bacterial DNA but cleaves the foreign DNA. But there are some restriction enzymes which function exactly in reverse mode. They cut the DNA if it is a methylated.

Star activity

Sometimes restriction enzymes recognize and cleave the DNA strand at the recognition site with asymmetrical palindromic sequence, for example Bam HI cuts at the sequence GA TCC, but under extreme conditions such as low ionic strength it will cleave in any of the following sequence NGA TCC, GPOA TCC, GGNTCC. Such an activity of the RE is called star activity.

Nomenclature of Restriction Enzymes

As a large number of restriction enzymes have been discovered, a uniform nomenclature system is adopted to avoid confusion. This nomenclature was first proposed by Smith and Nathans in 1973.

1) The first letter of the restriction enzymes (RE) should be from the first letter of the species name of the organism from which the enzyme is isolated. The letter should be written in capitals and italics, e.g. RE from *E coli* will have *E* as starting letter.

2) The second and the third letters of RE should be from the first and the second letter of the genus name of the organism. The letter should be written in lower case and should be in italics, e.g. RE from *E coli* will have *Eco* as starting words.

3) If the RE is isolated from the particular strain of an organism then that should be written as fourth letter. It should be in capitals and not in italics. For example, RE from *E coli* R strain will be written as *Eco* R.

If the RE isolated is the first of its kind from that particular organism, then the number I should be given. If already two REs are isolated, then number III should be given for the new restriction enzymes. The number should be written in roman, e.g. the first *E coli* RE should be written as *Eco* RI whereas the third restriction enzyme isolated from *E coli* R strains should be written as *Eco* RIII.

DNA LIGASE

It acts as a key player in recombinant DNA technology as a "molecular glue" or "molecular suture". The main source of this enzyme is the T4 phage. DNA ligase can join two different DNA pieces. The joining process requires ATP, as it needs energy to construct the bond. Normally the process is carried out at 4°C in order to reduce the kinetic energy of the molecule. This enzyme, when joining two DNA pieces, forms a covalent bond between the 5' phosphoryl of one strand and 3' hydroxyl of the adjacent strand. Hence it catalyses the end-to end joining of DNA duplex at the base-paired end. Mertz and Davis (1972) for the first time demonstrated that cohesive termini of cleaved DNA molecules could be covalently sealed with *E coli* DNA ligase and were able to produce recombinant DNA molecules.

The *E coli* DNA ligase uses nicotinamide adenine dinucleotide (NAD+) as a cofactor, while T4 DNA ligase requires ATP for the same. Both the enzymes contain an -NH2 group on the lysine residue. In both cases, the cofactor breaks into AMP (adenosine monophosphate), which in turn adenylates the enzyme (E) to form enzyme -AMP complex (EAC). EAC binds to the nick containing 3'OH and 5'PO_4 ends on a double-stranded DNA molecule. The 5' phosphoryl terminus of the nick is adenylated by the EAC with 3'OH terminus resulting in the formation of phosphodiester and liberation of AMP (Lehman, 1974).

After formation of **phosphodiester**, the nick is sealed. T4 DNA ligase has the ability to join the blunt ends of DNA fragments, whereas *E coli* DNA ligase joins the cohesive ends Produced by restriction enzymes. Additional advantage with T4 enzyme is that it can quickly join and produce the full base pairs but it would be difficult to retrieve the inserted DNA from vector. However cohesive end ligation proceeds about 100 times faster than the blunt end ligation.

KINASE

Kinase is the group of enzymes, which adds a free pyrophosphate (PO_4) to a wide variety of substrates like proteins, DNA and RNA. It uses ATP as cofactor and adds a phosphate by breaking the ATP into ADP and pyrophosphate. It is widely used in molecular biology and genetic engineering to add radiolabelled phosphates.

ALKALINE PHOSPHATASES

Alkaline phosphatase is a glycoprotein with two identical subunits. The cohesive ends of broken plasmids, instead of joining with foreign DNA, join the cohesive end of the same DNA molecules and get recircularized. To overcome this problem the restricted plasmid is treated with an enzyme, alkaline phosphatase that digests the terminal phosphoryl group. The restriction fragments of the foreign DNA to be cloned are not treated with alkaline phosphatase. Therefore, the 5' end of foreign DNA fragment can covalently join to 3' end of the plasmid. The recombinant DNA thus obtained has a nick with 3' and 5' P hydroxy ends. Ligase will only join 3' and 5' ends of recombinant DNA together if the 5' end is phosphorylated. Thus, alkaline phosphatase and ligase prevent recircularization of the vector and increase the frequency of production of recombinant DNA molecules. The nicks between two 3' ends fragment and vector DNA are repaired inside the bacterial cells during the transformation.

Phosphatases are a group of enzymes which remove a phosphate from a variety of substrates like DNA, RNA and proteins. Phosphatases which act in basic buffers with pH 8 or 9 are called as alkaline Phosphatases. Most commonly bacterial alkaline Phosphatases (BAP), calf intestine alkaline Phosphatases (CIAP) and shrimp alkaline Phosphatases are used in molecular cloning experiments.

The PO_4 from the substrate is removed by forming phosphorylated serine intermediate. Alkaline phosphatase are metalloenzymes, and have Zn++ ions in them. BP A (bacterial alkaline phosphatase) is a dimer containing six Zn++ ions, two of which are essential for enzymatic activity. BP A is very stable and is not inactivated by heat and detergent. Calf intestine alkaline phosphatases (CIAP) is also a dimer. It requires Zn++ and Mg++ ions for action. CIAP is inactivated by heating at 70°C for twenty min. or in the presence of 10 m MEGT A. Alkaline phosphatases are used to remove the PO_4 from the DNA or as reporter enzymes.

REVERSE TRANSCRIPTASE

Reverse transcriptase is used to synthesize the copy DNA or complementary DNA (using mRNA as a template). Reverse transcriptase is very essential in the synthesis of cDNA and construction of cDNA clone bank.

During the 1960s, Temin and his co-workers postulated that in certain cancer-causing animal viruses which contain RNA as genetic material, transcription of cancerous genes takes place most probably by DNA polymerase directed by virus RNA. Then DNA is used as template for synthesis of many copies of viral RNA in a cell Thus, they found that retroviruses (possessing RNA) contain RNA-dependent DNA polymerase, which is also called as reverse transcriptase. This produces single-stranded DNA, which in turn functions as template for complementary long chain of DNA. This enzyme uses an RNA molecule as template and synthesizes a DNA strand complementary to the RNA molecule.

These enzymes are used to synthesize the DNA from RNA. These enzymes are present in most of the RNA tumour viruses and retroviruses.

Reverse transcriptase enzyme is also called as RNA dependent DNA polymerase. Reverse transcriptase enzyme, after synthesizing the complementary strand at the 3 end of the DNA strand, adds a small extra nucleotide stretch without complementary sequence. This short stretch is called as R-loop.

TERMINAL DEOXYNUCLEOTIDE TRANSFERASE

Terminal deoxynucleotide Transferase is a polymerase which adds nucleotides at 3' -OH end (like klenow fragment) but does not require any complementary sequence and does not copy any DNA sequence (unlike klenow fragment). Terminal deoxynucleotide transferase (TDNT) adds nucleotide whatever comes into its active site and it does not show any preference for any nucleotide.

RNASE P

It specifically cleaves at the 5' end of RNA. It is a complex enzyme consisting of small protein (20 kilodaltons) and a 377 -nucleotide RNA molecule. It has been observed that the RNA molecule possesses at least part of the enzymatic activity of the complex. Hence, it is an example of ribozyme.

KLENOW FRAGMENT

E. coli DNA polymerase I consists of a single polypeptide chain. Pol I carries out three enzymatic reactions that are performed by three distinct functional domains. Two fragments are obtained when DNA pol I is treated with trypsin/subtilisin in mild conditions. The larger fragment is called as klenow fragment. This fragment is 602 amino acids in length. The function of the klenow fragment is to add nucleotides to the 3 end and 3 -5 exonuclease activity. Klenow fragment adds nucleotides by using complementary strand as reference. It cannot extend the DNA without the presence of the complementary strand.

If any nucleotide is added by mistake and the base pair is wrong (if A is paired to G instead of T) then by using 3 -5 exonuclease activity present in klenow fragment, this mispaired base pair is removed. In general the klenow fragment has 5 -3 polymerase and exonuclease activity.

A major step forward in gene modification was the discovery of restriction enzymes, which cleave DNA at specific sequences. These enzymes were discovered at approximately the same time as the first DNA ligases by Swiss biologist Werner Arber and his colleagues while they were investigating a phenomenon called host-controlled restriction of bacteriophages. Bacteriophages are viruses that invade and often destroy their bacterial host cells; host-controlled restriction refers to the defense mechanisms that bacterial cells have evolved to deal with these invading viruses. Arber's team discovered that one such mechanism is enzymatic activity provided by the host cell. The team named the responsible enzymes "restriction enzymes" because of the way they restrict the growth of bacteriophages. These scientists were also the first to demonstrate that restriction enzymes damage invading bacteriophages by cleaving the phage DNA at very specific nucleotide sequences (now known as restriction sites). The identification and characterization of restriction enzymes gave biologists the means to cut specific pieces of DNA required (or desired) for subsequent recombination.

3

Nucleic Acid Isolation

A **nucleic acid** is a macromolecule composed of chains of monomeric nucleotides. In biochemistry these molecules carry genetic information or form structures within cells. The most common nucleic acids are deoxyribonucleic acid (DNA) and ribonucleic acid (RNA). Nucleic acids are universal in living things, as they are found in all cells and viruses. Nucleic acids were first discovered by Friedrich Miescher in 1871. Artificial nucleic acids include peptide nucleic acid (PNA), Morpholino and locked nucleic acid (LNA), as well as glycol nucleic acid (GNA) and Threose nucleic acid (TNA). Each of these is distinguished from naturally-occurring DNA or RNA by changes to the backbone of the molecule.

DEOXYRIBONUCLEIC ACID

Deoxyribonucleic acid contains the genetic instructions used in the development and functioning of all known living organisms. The main role of DNA molecules is the long-term storage of information and DNA is often compared to a set of blueprints, since it contains the instructions needed to construct other components of cells, such as proteins and RNA molecules. The DNA segments that carry this genetic information are called genes, but other DNA sequences have structural purposes, or are involved in regulating the use of this genetic information.

DNA is made of four types of nucleotides, containing different nucleobases: the pyrimidines cytosine and thymine, and the purines guanine and adenine. The nucleotides are attached to each other in a chain by bonds between their sugar and phosphate groups, forming a sugar-phosphate backbone. Two of these chains are held together by hydrogen bonding between complementary bases; the chains coil around each other, forming the DNA double helix.

RIBONUCLEIC ACID

Ribonucleic acid, or RNA, is a nucleic acid polymer consisting of nucleotide monomers, which plays several important roles in the processes of transcribing genetic information from deoxyribonucleic acid (DNA) into proteins. RNA acts as a messenger between DNA and the protein synthesis complexes known as ribosomes, forms vital portions of ribosomes, and serves as an essential carrier molecule for amino acids to be used in protein synthesis. The three types of RNA include tRNA (transfer), mRNA (messenger) and rRNA (ribosomal).

DNA ISOLATION

Many different methods and technologies are available for the isolation of genomic DNA. In general, all methods involve disruption and lysis of the starting material followed by the removal of proteins and other contaminants and finally recovery of the DNA. Removal of proteins is typically achieved by digestion with proteinase K, followed by salting-out, organic extraction, or binding of the DNA to a solid-phase support (either anion-exchange or silica technology). DNA is usually recovered by precipitation using ethanol or isopropanol. The choice of a method depends on many factors: the required quantity and molecular weight of the DNA, the purity required for downstream applications, and the time and expense. Several of the most commonly used methods are detailed below, although many different methods and variations on these methods exist. Home-made methods often work well for researchers who have developed and regularly use them. However, they usually lack standardization and therefore yields and quality are not always reproducible. Reproducibility is also affected when the method is used by different researchers, or with different sample types. The separation of DNA from cellular components can be divided into four stages:

1) Disruption
2) Lysis
3) Removal of proteins and contaminants
4) Recovery of DNA

In some methods, stages 1 and 2 are combined.

Preparation of Crude Lysate

An easy technique for isolation of genomic DNA is to incubate cell lysates at high temperatures (e.g., 90°C for 20 minutes), or to perform a proteinase K digestion, and then use the lysate directly in downstream applications. Considered "quick-and-dirty" techniques, these methods are only appropriate for a limited range of applications. The treated lysate usually contains enzyme-inhibiting contaminants, such as salts, and DNA is often not at optimal pH. Furthermore, incomplete inactivation of proteinase K can result in false negative results and high failure rates. It is not recommended to store DNA prepared using this method, as the high levels of contamination often result in DNA degradation.

Salting-out Methods

Starting with a crude lysate, "salting-out" is another conventional technique where proteins and other contaminants are precipitated from the cell lysate using high concentrations of salt such as potassium acetate or ammonium acetate. The precipitates are removed by centrifugation, and the DNA is recovered by alcohol precipitation. Removal of proteins and other contaminants using this method may be inefficient, and RNase treatment, dialysis, and/or repeated alcohol precipitation are often necessary before the DNA can be used in downstream applications. DNA yield and purity are highly variable using this method.

Organic Extraction Methods

Organic extraction is a conventional technique that uses organic solvents to extract contaminants from cell lysate **(Fig 3.1)**. The cells are lysed using a detergent, and then mixed with phenol, chloroform, and isoamyl alcohol. The correct salt concentration and pH must be used during extraction to ensure that contaminants are separated into the organic phase and that DNA remains in the aqueous phase. DNA is usually recovered from the aqueous phase by alcohol precipitation. This is a time-consuming and cumbersome technique.

Furthermore, the procedure uses toxic compounds and may not give reproducible yields. DNA isolated using this method may contain residual phenol and/or chloroform, which can inhibit enzyme reactions in downstream applications, and therefore may not be sufficiently pure for sensitive downstream applications such as PCR. The process also generates toxic waste that must be disposed of with care and in accordance with hazardous waste guidelines. In addition, this technique is almost impossible to automate, making it unsuitable for high-throughput applications.

Cesium Chloride Density Gradients

Genomic DNA can be purified by centrifugation through a cesium chloride (CsCl) density gradient. Cells are lysed using a detergent, and the lysate is alcohol precipitated. Resuspended DNA is mixed with CsCl and ethidium bromide and centrifuged for several hours. The DNA band is collected from the centrifuge tube, extracted with isopropanol to remove the ethidium bromide, and then precipitated with ethanol to recover the DNA. This method allows the isolation of high-quality DNA, but is time consuming, labor intensive, and expensive (an ultracentrifuge is required), making it inappropriate for routine use. This method uses toxic chemicals and is also impossible to automate.

Anion-exchange Methods

Solid-phase anion-exchange chromatography is based on the interaction between the negatively charged phosphates of the nucleic acid and positively charged surface molecules on the substrate. DNA binds to the substrate under low-salt conditions, impurities such as RNA, cellular proteins, and metabolites are washed away using medium salt buffers and high-quality DNA is eluted using a high-salt buffer. The eluted DNA is recovered by alcohol precipitation, and is suitable for all downstream applications. Anion-exchange technology completely avoids the use of toxic substances, and can be used for different throughput requirements as well as for different scales of purification. The isolated DNA is sized up to 150 kb, with an average length of 50–100 kb.

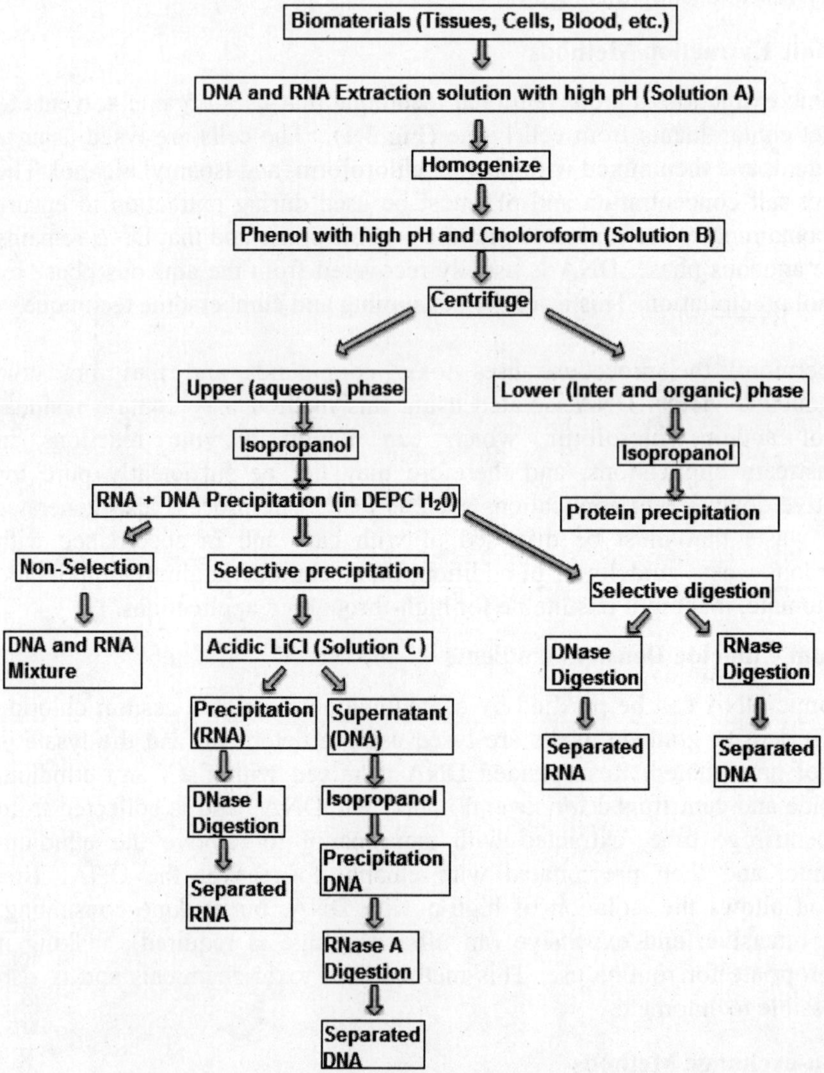

Fig. 3.1: Nucleic acid isolation

Silica-based Methods

DNeasy Tissue technology provides a simple, reliable, fast, and inexpensive method for isolation of high-quality DNA. This method is based on the selective adsorption of nucleic acids to a silica-gel membrane in the presence of high concentrations of chaotropic salts. Use of optimized buffers in the lysis procedure ensures that only DNA is adsorbed while cellular proteins and metabolites remain in solution and are subsequently washed away. This is simpler and more effective than other methods where precipitation or extraction is required. Ready-to-use DNA is then eluted from the silica-gel membrane using a low-salt buffer. No alcohol precipitation is required, and resuspension of the DNA, which is often difficult if the DNA has been over-dried, is not required.

RNA ISOLATION

Obtaining high quality, intact RNA is the first and often the most critical step in performing many fundamental molecular biology experiments, including Northern analysis, nuclease protection assays, RT-PCR, RNA mapping, in vitro translation and cDNA library construction. To be successful, however, the RNA isolation procedure should include some important steps both before and after the actual RNA purification.

During tissue disruption for RNA isolation, it is crucial that the denaturant be in contact with the cellular contents at the moment that the cells are disrupted. This can be problematic when tissues/cells are hard (e.g. bone, roots), when they contain capsules or walls (e.g. yeast, Gram-positive bacteria) or, when samples are numerous, making rapid processing difficult. A common solution to these problems is to freeze the tissue/cells in liquid nitrogen or on dry ice. The samples must then be ground with a mortar and pestle into a fine powder, which is added to the denaturant. While this freezing and grinding process allows the researcher to postpone RNA isolation, it is a time consuming and laborious process.

Tissue or Cell Sample Collection and Disruption

Ongoing research into optimizing RNA analysis has identified two points in the RNA isolation process that can be improved; treatment and handling of tissue or cells prior to RNA isolation and storage of the isolated RNA. Since most of the actual RNA isolation procedure takes place in a strong denaturant (e.g. GITC, LiCl, SDS, phenol) that renders RNases inactive, it is typically prior to, and after the isolation, when RNA integrity is at risk. Finding the most appropriate method of cell or tissue disruption for specific starting material is important for maximizing the yield and quality of RNA preparation.

Cell Disruption: Getting the RNA Out

Complete Disruption - A Critical Step

Cellular disruption is the first step in RNA isolation and one of the most critical steps affecting yield and quality of the isolated RNA. Typically, cell disruption needs to be fast and thorough. Slow disruption, for example placing cells or tissue in guanidinium isothiocyanate (GITC) lysis solution without any additional physical shearing, may result in RNA degradation by endogenous RNases released internally, yet still inaccessible to the protein denaturant, GITC. This is especially a concern when working with tissues high in endogenous RNase such as spleen and pancreas. Incomplete disruption may also result in decreased yield because some of the RNA in the sample remains trapped in intact cells and, therefore, is unavailable for subsequent purification. For most samples, thorough disruption can be monitored by close inspection of the lysate after disruption. There should be no visible particulates, except when disrupting materials containing hard, non-cellular components, such as connective tissue or bone.

Mechanical or Enzymatic Disruption

Mechanical

Cell and tissue disruption methods are usually mechanical or enzymatic. Mechanical methods for disrupting fresh tissue and cells include homogenization with a Dounce or with a mechanical homogenizer (such as the Brinkmann Polytron), vortexing, sonication, French press, bead milling, and even grinding in a coffee grinder! Disrupting frozen tissue is more time consuming and cumbersome than processing fresh tissue, but freezing tissue samples is sometimes necessary. Samples are usually frozen when, 1) they are collected over a period of time and thus, cannot be processed simultaneously; 2) there are many samples, 3) samples are collected in the field, or 4) mechanical processing of fresh samples is insufficient for thorough disruption. A mortar and pestle or bag and hammer are typically used when the starting material is frozen. RNA will remain intact in tissues for a day at 37°C, a week at 25°C, a month at 4°C, and indefinitely at subzero temperatures.

Enzymatic

Lysozyme and zymolase digestion are among the enzymatic methods frequently used with bacteria and yeast to dissolve a coat, capsule, capsid or other structure not easily sheared by mechanical methods alone. Enzymatic treatment is usually followed by sonication, homogenization or vigorous vortexing in a GITC lysis buffer. Enzymatic methods may also be used for specific eukaryotic tissues, e.g., collagenase to break down collagen prior to cell lysis.

Plant Tissues

Soft, fresh plant tissues from plants such as *Nicotiana* and *Arabidopsis* can often be disrupted by homogenization in lysis buffer alone. Other plant tissues, like pine needles, need to be ground dry, without liquid nitrogen. Some hard, woody plant materials may require freezing and grinding in liquid nitrogen or milling. Plant cell suspension cultures and calluses can be lysed by sonication in a lysis buffer for 0.5–2 minutes. The diversity of plants and plant tissue make it impossible to give a single recommendation for all. However, most plant tissues typically contain polysaccharides and polyphenols that can coprecipitate with RNA and inhibit downstream assays. Treating a plant tissue lysate with polyvinylpyrrolidone (PVP) will precipitate such problematic components from the lysate before the actual RNA isolation is carried out.

Yeast and Fungi

Yeast can be extremely difficult to disrupt because their cell walls may form capsules or nearly indestructible spores. There are several ways to approach yeast cell disruption. One of the most common and probably the most straightforward methods is mechanical disruption using a bead mill. Bead mills vigorously agitate a tube containing the sample, lysis buffer and small glass beads (0.5–1 mm).

Alternatively, yeast cell walls can be digested with zymolase, glucalase and/or lyticase to produce spheroplasts that are readily lysed by vortexing in a guanidinium-based lysis buffer. Some specialized isolation methods for yeast exist which use such methods as boiling SDS or boiling phenol treatment to insure complete cell lysis.

To disrupt filamentous fungi, scrape the mycelial mat into a cold mortar, add liquid nitrogen and grind to a fine powder with a pestle. The powder can then be thoroughly homogenized or sonicated in lysis buffer to solubilize completely. As fungi may also be rich in polysaccharides, treatment with PVP may be helpful here too.

Bacteria

Bacteria, like plants, are extremely diverse; therefore, it is difficult to make one recommendation for all bacteria. Bead milling will lyse most Gram positive and Gram negative bacteria, including mycobacteria. It can be performed by adding glass beads and lysis solution to a bacterial cell pellet and milling for a few minutes. It is possible to lyse some Gram negative bacteria by sonication in lysis solution alone. It has been found this to be sufficient for small cultures (milliliters), but not large cultures (liters). Bacterial cell walls can be digested with lysozyme to form spheroplasts. Gram positive bacteria usually require more rigorous digestion (increased incubation time, increased incubation temperature, etc.) than Gram negative organisms. The spheroplasts are then easily lysed with vigorous vortexing or sonication in GITC lysis buffer.

Soil and Sediments

Disruption of cells found in soil and sediments is accomplished one of two ways. One technique isolates the bacterial cells from the material prior to the RNA isolation procedure. This is accomplished by homogenization of wet soil in a Waring blender followed by a slow speed centrifugation to remove fungal biomass and soil debris. The supernatant is centrifuged again at a higher speed to pellet the bacterial cells (5). From this point, cells can be lysed as described above for bacteria. Other techniques describe RNA isolation from the soil or sediment directly. For example, one method requires soil to be added to a bead mill along with diatomaceous earth and lysis buffer. The sample is then agitated for a few minutes and centrifuged to remove solid debris.

Eukaryotic mRNA

Poly (A) RNA (mRNA) makes up between 1-5% of total cellular RNA and is most frequently used for 1) detection and quantitation of extremely rare mRNAs, 2) synthesis of probes for array analysis, and 3) the construction of random-primed cDNA libraries, where the use of total RNA would generate rRNA templates that would significantly dilute out cDNAs of interest. Removal of ribosomal and transfer RNA results in up to a 30-fold enrichment of a specific message.

Prokaryotic mRNA

For decades mRNA has been isolated from eukaryotic sources using oligo(dT) selection. Bacteria, however, lack the relatively stable poly(A) tails found on eukaryotic messages. Until very recently, isolating mRNA from bacteria has been virtually impossible.

Speed is critical in the purification of bacterial RNA due to the short half-life of bacterial mRNA and the need to rapidly "freeze" the mRNA expression profile. Some bacterial isolation protocols call for the pretreatment of bacteria with lytic enzymes. While this pretreatment does assist lysis, it delays isolation and may lead to altered expression profiles.

The following methods are better alternatives for effectively freezing gene expression profiles:

- Immediate cell lysis and RNA purification
- Rapid freezing in liquid nitrogen (a freeze-thaw treatment may help with lysis of some bacteria)

Easily lysed Gram-negative bacteria may be pipetted directly into a boiling lysis buffer of choice (without even removing the culture medium), and RNA can be immediately extracted. Most other bacteria will need to be pelleted by brief centrifugation prior to the above treatments.

To purify RNA from a bacterial cell pellet, add boiling lysis buffer to the pellet and vortex rapidly, then immediately extract the lysate with hot acid phenol: chloroform. Harsh mechanical devices (e.g. bead mills) may be required to disrupt some bacterial species. Once lysed, extract the preparation with hot acid phenol: chloroform Reagent. Alternatively, it may be possible to disrupt bacteria directly in acid phenol: chloroform using a bead mil.

Storage of Isolated RNA

The last step in every RNA isolation protocol, whether for total or mRNA preparation, is to resuspend the purified RNA pellet. After painstakingly preparing an RNA sample, it is crucial that RNA be suspended and stored in a safe, RNase-free solution.

For long term storage, RNA samples may also be stored at -20°C as ethanol precipitates. Accessing these samples on a routine basis can be a nuisance, however, since the precipitates must be pelleted and dissolved in an aqueous buffer before pipetting, if accurate quantitation is important. An alternative is to pipette directly out of an ethanol precipitate that has been vortexed to create an even suspension. It is found, however, that while this method is suitable for qualitative work, it is too imprecise for use in quantitative experiments. RNA does not disperse uniformly in ethanol, probably because it forms aggregates; non-uniform suspension, in turn, leads to inconsistency in the amount of RNA removed when equal volumes are pipetted.

QUANTIFICATION OF NUCLEIC ACIDS

Quantification of nucleic acids is commonly used in molecular biology to determine the concentrations of DNA or RNA present in a mixture, as subsequent reactions or protocols using a nucleic acid sample often require particular amounts for optimum performance. There exist several methods to establish the concentration of a solution of nucleic acids, including spectrophotometric quantification and UV fluorescence in presence of a DNA dye.

Spectrophotometric Quantification

Because DNA and RNA absorb ultraviolet light, with an absorption peak at 260nm wavelength, spectrophotometers are commonly used to determine the concentration of DNA in a solution. Inside a spectrophotometer, a sample is exposed to ultraviolet light at 260 nm, and a photo-detector measures the light that passes through the sample. The more light absorbed by the sample, the higher the nucleic acid concentration in the sample.

Using the Beer Lambert Law it is possible to relate the amount of light absorbed to the concentration of the absorbing molecule. At a wavelength of 260 nm, the average extinction coefficient for double-stranded DNA is 0.020

$(\mu g/ml)^{-1}$ cm^{-1}, for single-stranded DNA and RNA it is 0.027 $(\mu g/ml)^{-1}$ cm^{-1} and for short single-stranded oligonucleotides it is dependent on the length and base composition. Thus, an optical density (or "OD") of 1 corresponds to a concentration of 50 µg/ml for double-stranded DNA. This method of calculation is valid for up to an OD of at least 2. A more accurate extinction coefficient may be needed for oligonucleotides; these can be predicted using the nearest-neighbor model.

Sample purity

It is common for nucleic acid samples to be contaminated with other molecules (eg, protein, phenol, and other organic compounds). Because these molecules have their own characteristic absorption spectra, the absorption at other wavelengths is often compared to 260nm absorption in order to assess sample purity. In addition, some contaminants (notably phenol) can significantly contribute to an error in concentration estimation as they also absorb strongly at 260nm.

Protein contamination and the 260:280 ratio

The ratio of absorptions at 260 nm vs. 280 nm is commonly used to assess the purity of protein with respect to DNA contamination, since protein (in particular, the aromatic amino acids) tends to absorb at 280nm. The method dates back to 1942, when Warburg and Christian showed that the ratio is a good indicator of nucleic acid contamination in protein preparations. The reverse, however, is not true—it takes a relatively large amount of protein contamination to significantly affect the 260:280 ratios.

260:280 ratios have high sensitivity for nucleic acid contamination in protein:

% Protein	% Nucleic acid	260:280 ratio
100	0	0.57
95	5	1.06
90	10	1.32
70	30	1.73

260:280 ratio lacks sensitivity for protein contamination in nucleic acids:

% nucleic acid	% protein	260:280 ratio
100	0	2.00
95	5	1.99
90	10	1.98
70	30	1.94

This difference is due to the much higher extinction coefficient nucleic acids have at 260nm and 280nm, compared to that of proteins. Because of this, even for relatively high concentrations of protein, the protein contributes relatively little to the 260 and 280 absorbance. While the protein contamination cannot be reliably assessed with a 260:280 ratio, this also means that it contributes little error to DNA quantity estimation.

Other common contaminants

Contamination by phenol, which is commonly used in nucleic acid purification, can significantly throw off quantification estimates. Phenol absorbs with a peak at 270nm and a 260:280 ratio of 2. Nucleic acid preparations uncontaminated by phenol should have a 260:270 ratio of around 1.2. Contamination by phenol can significantly contribute to overestimation of DNA concentration.

Absorption at 230nm can be caused by contamination by phenolate ion, thiocyanates, and other organic compounds. For a pure RNA sample, the 260:230 ratio should be around 2, and for a pure DNA sample, the 280:230 ratio should be around 1.8.

Absorption at 330nm and higher indicates particulates contaminating the solution, causing scattering of light in the visible range. The value in a pure nucleic acid sample should be zero.

Negative values could result if an incorrect solution was used as blank. Alternatively, these values could arise due to fluorescence of a dye in the solution.

Quantification using Fluorescent Dyes

An alternative way to assess DNA concentration is to use measure the fluorescence intensity of dyes that bind to nucleic acids and selectively fluoresce when bound (e.g. Ethidium bromide). This method is useful for cases where concentration is too low to accurately assess with spectrophotometry and in cases where contaminants absorbing at 260nm make accurate quantitation by that method impossible.

There are two main ways to approach this. "Spotting" involves placing a sample directly onto an agarose gel or plastic wrap. The fluorescent dye is either present in the agarose gel, or is added in appropriate concentrations to the samples on the plastic film. A set of samples with known concentrations are spotted alongside the sample. The concentration of the unknown sample is then estimated by comparison with the fluorescence of these known concentrations. Alternatively, one may run the sample through an agarose or polyacrylamide gel, alongside some samples of known concentration. As with the spot test, concentration is estimated through comparison of fluorescent intensity with the known samples.

If the sample volumes are large enough to use microplates or cuvettes, the dye-loaded samples can also be quantified with a fluorescence photometer.

4

Polymerase Chain Reaction

The polymerase chain reaction (PCR) is a technique widely used in molecular biology, microbiology, genetics, diagnostics, clinical laboratories, forensic science, environmental science, hereditary studies, paternity testing, and many other applications. The name, polymerase chain reaction, comes from the DNA polymerase used to amplify (replicate many times) a piece of DNA by *in vitro* enzymatic replication. The original molecule or molecules of DNA are replicated by the DNA polymerase enzyme, thus doubling the number of DNA molecules. Then each of these molecules is replicated in a second "cycle" of replication, resulting in four times the number of the original molecules. Again, each of these molecules is replicated in a third cycle of replication. This process is known as a "chain reaction" in which the original DNA template is exponentially amplified. With PCR it is possible to amplify a single piece of DNA, or a very small number of pieces of DNA, over many cycles, generating millions of copies of the original DNA molecule. PCR has been extensively modified to perform a wide array of genetic manipulations, diagnostic tests, and for many other uses.

The polymerase chain reaction is used by a wide spectrum of scientists in an ever-increasing range of scientific disciplines. In microbiology and molecular biology, for example, PCR is used in research laboratories in DNA cloning procedures, Southern blotting, DNA sequencing, recombinant DNA technology, to name but a few. In clinical microbiology laboratories PCR is invaluable for the diagnosis of microbial infections and epidemiological studies. PCR is also used in forensics laboratories and is especially useful because only a tiny amount of original DNA is required, for example, sufficient DNA can be obtained from a droplet of blood or a single hair.

In molecular biology, the **polymerase chain reaction (PCR)** is a technique to amplify a single or few copies of a piece of DNA across several orders of magnitude, generating thousands to millions of copies of a particular DNA sequence. The method relies on thermal cycling, consisting of cycles of repeated heating and cooling of the reaction for DNA melting and enzymatic replication of the DNA. Primers (short DNA fragments) containing sequences complementary to the target region along with a DNA polymerase (after which the method is named) are key components to enable selective and repeated amplification. As PCR progresses, the DNA generated is itself used as a template for replication, setting in motion a chain reaction in which the DNA template is exponentially amplified. PCR can be extensively modified to perform a wide array of genetic manipulations.

Almost all PCR applications employ a heat-stable DNA polymerase, such as *Taq* polymerase, an enzyme originally isolated from the bacterium *Thermus aquaticus*. This DNA polymerase enzymatically assembles a new DNA strand from DNA building blocks, the nucleotides, by using single-stranded DNA as a template and DNA oligonucleotides (also called DNA primers), which are required for initiation of DNA synthesis. The vast majority of PCR methods use thermal cycling, i.e., alternately heating and cooling the PCR sample to a defined series of temperature steps. These thermal cycling steps are necessary first to physically separate the two strands in a DNA double helix at a high temperature in a process called DNA melting. At a lower temperature, each strand is then used as the template in DNA synthesis by the DNA polymerase to selectively amplify the target DNA. The selectivity of PCR results from the use of primers that are complementary to the DNA region targeted for amplification under specific thermal cycling conditions.

Developed in 1983 by Kary Mullis, PCR is now a common and often indispensable technique used in medical and biological research labs for a variety of applications. These include DNA cloning for sequencing, DNA-based phylogeny, or functional analysis of genes; the diagnosis of hereditary diseases; the identification of genetic fingerprints (used in forensic sciences and paternity testing); and the detection and diagnosis of infectious diseases. In 1993, Mullis was awarded the Nobel Prize in Chemistry for his work on PCR.

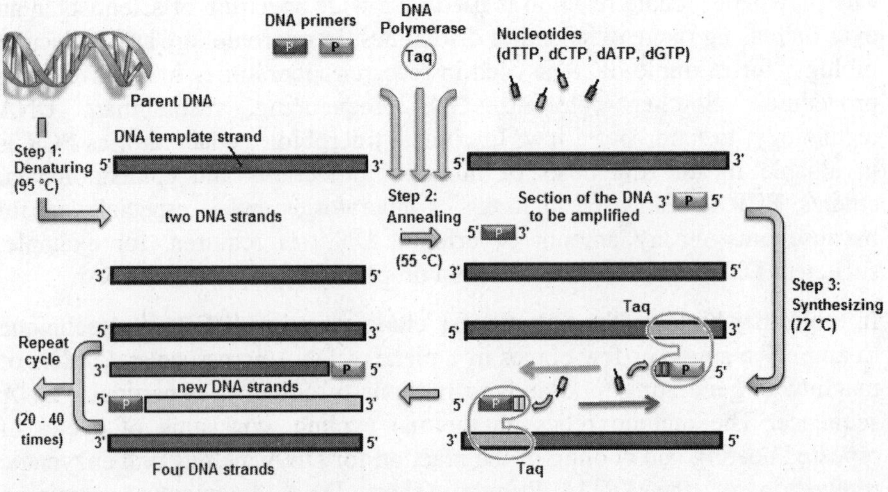

Fig. 4.1: Polymerase Chain Reaction

PROCEDURE

The PCR usually consists of a series of 20-40 repeated temperature changes called cycles; each cycle typically consists of 2-3 discrete temperature steps. Most commonly PCR is carried out with cycles that have three temperature steps. The cycling is often preceded by a single temperature step (called *hold*) at a high temperature (>90°C), and followed by one hold at the end for final product extension or brief storage. The temperatures used and the length of time they are applied in each cycle depend on a variety of parameters (Fig. 4.1). These include the enzyme used for DNA synthesis, the concentration of divalent ions and dNTPs in the reaction, and the melting temperature (Tm) of the primers.

- **Initialization step:** This step consists of heating the reaction to a temperature of 94–96 °C (or 98 °C if extremely thermostable polymerases are used), which is held for 1–9 minutes. It is only required for DNA polymerases that require heat activation by hot-start PCR.

- **Denaturation step:** This step is the first regular cycling event and consists of heating the reaction to 94–98 °C for 20–30 seconds. It causes DNA melting of the DNA template by disrupting the hydrogen bonds between complementary bases, yielding single strands of DNA.

- **Annealing step:** The reaction temperature is lowered to 50–65 °C for 20–40 seconds allowing annealing of the primers to the single-stranded DNA template. Typically the annealing temperature is about 3-5 degrees Celsius below the Tm of the primers used. Stable DNA-DNA hydrogen bonds are only formed when the primer sequence very closely matches the template sequence. The polymerase binds to the primer-template hybrid and begins DNA synthesis.

- **Extension/elongation step:** The temperature at this step depends on the DNA polymerase used; Taq polymerase has its optimum activity temperature at 75–80 °C, and commonly a temperature of 72 °C is used with this enzyme. At this step the DNA polymerase synthesizes a new DNA strand complementary to the DNA template strand by adding dNTPs that are complementary to the template in 5' to 3' direction, condensing the 5'-phosphate group of the dNTPs with the 3'-hydroxyl group at the end of the nascent (extending) DNA strand. The extension time depends both on the DNA polymerase used and on the length of the DNA fragment to be amplified. As a rule-of-thumb, at its optimum temperature, the DNA polymerase will polymerize a thousand bases per minute. Under optimum conditions, i.e., if there are no limitations due to limiting substrates or reagents, at each extension step, the amount of DNA target is doubled, leading to exponential (geometric) amplification of the specific DNA fragment.

- **Final elongation:** This single step is occasionally performed at a temperature of 70–74 °C for 5–15 minutes after the last PCR cycle to ensure that any remaining single-stranded DNA is fully extended.
- **Final hold:** This step at 4–15 °C for an indefinite time may be employed for short-term storage of the reaction.

To check whether the PCR generated the anticipated DNA fragment (also sometimes referred to as the amplimer or amplicon), agarose gel electrophoresis is employed for size separation of the PCR products. The size(s) of PCR products is determined by comparison with a DNA ladder (a molecular weight marker), which contains DNA fragments of known size, run on the gel alongside the PCR products.

The first extension products result from DNA synthesis on the original template and these do not have a distinct length as the DNA polymerase will continue to synthesis new DNA until it either stops or is interrupted by the start of the next cycle. The second cycle extension products are also of intermediate length, at the 3rd cycle, fragments of "target" sequence are synthesized which are of defined length corresponding to the position of the primers on the original template. From the 4th cycle onwards, the target sequence is amplified exponentially. This amplification, as a final number of copies of the target sequence, is expressed by the formula,

(2n–2n)X, where
n = number of cycles
2n = first product obtained after cycle 1 and second product obtained after cycle 2 with undefined length
X = number of copies of the original template

Potentially after 20 cycles of PCR, there will be 220-fold amplification, assuming 100% efficiency during each cycle. However, in practice, only 20-30% efficiency is achieved in PCR methods.

The original PCR protocols used Klenow fragment of *E. coli* DNA polymerase I to catalyse the oligonucleotide extension. However, this enzyme is thermally inactivated during the denaturation step of a PCR cycle and so the researchers had to add a fresh aliquot of enzyme at each cycle, to the amplification process. In later years, a thermostable polymerase like Taq polymerase was discovered for use in PCR methods.

DETECTION AND ANALYSIS OF PCR PRODUCTS

The PCR products or amplicons consists of a fragment (or fragments) of DNA which is normally of a length defined by the boundaries of the PCR primers. PCR products are generally less than 10 kb in length. Many techniques can be used to detect and confirm the identity of the amplified products. Normally, the simplest or most commonly used method is electrophoresis of an aliquot of the PCR product on an agarose or polyacrylamide gel and visualization by staining with ethidium bromide, which is a fluorescent dye that intercalates into the DNA. After staining, ultraviolet transillumination allows visualization of the DNA in the gel.

Detection and confirmation of identity of a PCR product can be performed by Southern blot hybridization or dot blot hybridization and detection with radioactive or non-radioactive specific probes. The Southern blot techniques involve capillary blotting of the agarose or polyacrylamide gel onto a nitrocellulose or nylon membrane. The filter is hybridized with probe labelled with ^{32}P or non-radioactive label like digoxigenin label.

The PCR product can also be digested with restriction enzyme and the product is run on agarose gel for further characterization.

MAJOR ADVANTAGES OF PCR

Because of its simplicity, PCR is a popular technique with a wide range of applications which depend on essentially three major advantages of the method.

Speed and Ease of Use

DNA cloning by PCR can be performed in a few hours, using relatively unsophisticated equipment. Typically, a PCR reaction consists of 30 cycles containing a denaturation, synthesis and re-annealing step, with an individual cycle typically taking 3-5 minutes in an automated thermal cycler.

This compares favorably with the time required for cell-based DNA cloning, which may take weeks. Clearly, some time is also required for designing and synthesizing oligonucleotide primers, but this has been simplified by the availability of computer software for primer design and rapid commercial synthesis of custom oligonucleotides. Once the conditions for a reaction have been tested, the reaction can be repeated simply.

Comparison PCR - Polymerase Chain Reaction and Gene Cloning

	Parameter	PCR	Gene Cloning
1	Final Result	Selective amplification of specific sequence	Selective amplification of specific sequence
2	Manipulation	In vitro	In vitro and in vivo
3	Selectivity of the specific segment from complex DNA	First step	Last step

4	Quantity of starting material	Nanogram (ng)	Microgram (m)
5	Biological reagents required	DNA polymerase (Taq polymerase)	Restriction enzymes, Ligase, vector, bacteria
6	Automation	Yes	No
7	Labour intensive	No	Yes
8	Error probability	Less	More
9	Applications	More	Less
10	Cost	Less	More
11	User's skill	Not required	Required
12	Time of typical experiment	Four hours	Two to four days

Sensitivity

PCR is capable of amplifying sequences from minute amounts of target DNA, even the DNA from a single cell. Such exquisite sensitivity has afforded new methods of studying molecular pathogenesis and has found numerous applications in forensic science, in diagnosis, in genetic linkage analysis using single-sperm typing and in molecular palaentology studies, where samples may contain minute numbers of cells. However, the extreme sensitivity of the method means that great care has to be taken to avoid contamination f the sample under investigation by external DNA, such as from minute amounts of cells from the operator.

Robustness

PCR can permit amplification of specific sequences from material in which the DNA is badly degraded or embedded in a medium from which conventional DNA isolation is problematic. As a result, it is again very suitable for molecular anthropological and palaentological studies, for example, the analysis of DNA recovered from archaeological remains. It has also been used successfully to amplify DNA from formalin-fixed tissue samples, which has important applications in molecular pathology and, in some cases, genetic linkage studies.

THERMOSTABLE DNA POLYMERASES AND THEIR SOURCES

DNA polymerase	Natural/recombinant	Source
Taq	Natural	*Thermus aquaticus*
AmplitaqR	Recombinant	*T. aquaticus*
Hot TubTM	Natural	*Thermus flavus*
VentTM	Recombinant	*Thermococcus litoralis*

Tth	Recombinant	*Thermus thermophilus*
Pfu	Natural	*Pyrococcus furiosus*
UITmaTM	Recombinant	*Thermotoga maritima*

Vent TM DNA Polymerase

Vent TM DNA polymerase was first isolated from *Thermococcus litoralis*, which is a thermophilic archaebacterium found on ocean floors at temperature of up of 98ºC. The gene encoding this enzyme has been cloned and expressed in E. coli (New England Biolabs). This polymerase is more thermostable than Taq DNA polymerase and is capable of extending primers to give products up to 8-13 kb in length. Vent TM polymerase possesses a 3' to 5' exonuclease activity that is responsible for the high level of fidelity, which is 5 to 15-fold greater than that of Taq DNA polymerase.

Pfu DNA Polymerase

This polymerse isolated from the hyperthermophilic marine archaebacterium *Pyrococcus furiousus*, possesses both 5' to 3' DNA polymerase activity and 3' and 5' exonuclease proof-reading activity. The fidelity of DNA synthesis is 12-fold higher than that of Taq DNA polymerase. Pfu DNA polymerase is available commercially from Stratagene, as a genetically engineered mutant of cloned Pfu DNA polymerase. When setting up reaction with this enzyme, it is essential to add the enzyme last (as for Vent TM DNA polymerase), since in the absence of dNTPs the 3' to 5' exonuclease activity of the enzyme results in the degradation of templates and primers.

TYPES OF POLYMERASE CHAIN REACTION

Reverse Transcribed PCR

RT-PCR (reverse transcription-polymerase chain reaction) is the most sensitive technique for mRNA detection and quantitation currently available. Compared to the two other commonly used techniques for quantifying mRNA levels, Northern blot analysis and RNase protection assay, RT-PCR can be used to quantify mRNA levels from much smaller samples. In fact, this technique is sensitive enough to enable quantitation of RNA from a single cell.

RT-PCR utilizes a pair of primers, which are complementary to a defined sequence on each of the two strands of the cDNA. These primers are then extended by a DNA polymerase and a copy of the strand is made after each cycle, leading to logarithmic amplification.

RT-PCR includes three major steps. The first step is the reverse transcription (RT) where RNA is reverse transcribed to cDNA using a reverse transcriptase and primers. This step is very important in order to allow the performance of PCR since DNA polymerase can act only on DNA templates. The RT step can be performed either in the same tube with PCR (one-step PCR) or in a separate one (two-step PCR) using a temperature between 40°C and 50°C, depending on the properties of the reverse transcriptase used.

The next step involves the denaturation of the dsDNA at 95°C, so that the two strands separate and the primers can bind again at lower temperatures and begin a new chain reaction. Then, the temperature is decreased until it reaches the annealing temperature which can vary depending on the set of primers used, their concentration, the probe and its concentration (if used), and the cations concentration. The main consideration, of course, when choosing the optimal annealing temperature is the melting temperature (Tm) of the primers and probes (if used). The annealing temperature chosen for a PCR depends directly on length and composition of the primers. This is the result of the difference of hydrogen bonds between A-T (2 bonds) and G-C (3 bonds). An annealing temperature about 5 degrees below the lowest Tm of the pair of primers is usually used.

The final step of PCR amplification is the DNA extension from the primers which is done by the thermostable Taq DNA polymerase usually at 72°C, which is the optimal temperature for the polymerase to work. The length of the incubation at each temperature, the temperature alterations and the number of cycles are controlled by a programmable thermal cycler. The analysis of the PCR products depends on the type of PCR applied. If a conventional PCR is used, the PCR product is detected using agarose gel electrophoresis and ethidium bromide (or other nucleic acid staining).

Conventional RT-PCR is a time-consuming technique with important limitations when compared to real time PCR techniques. This, combined with the fact that ethidium bromide has low sensitivity, yields results that are not always reliable. Moreover, there is an increased cross-contamination risk of the samples since detection of the PCR product requires the post-amplification processing of the samples. Furthermore, the specificity of the assay is mainly determined by the primers, which can give false-positive results. However, the most important issue concerning conventional RT-PCR is the fact that it is a semi or even a low quantitative technique, where the amplicon can be visualized only after the amplification ends.

Inverse PCR

In this technique, the amplification of those DNA sequences takes place, which are away from the primers and not those which are flanked by the primers.

For instance if the border sequences of a DNA segment are not known and those of a vector are known, then the sequence to be amplified may be cloned in the vector and border sequences of vector may be used as primers in such a way that the polymerization proceeds in inverse direction i.e. away from the vector sequence flanked by the primers and towards the DNA sequence of inserted segment.

Similarly, if the gene sequence is known, one can use its border sequences as primers, for an inverse PCR reaction, to amplify the sequences flanking this gene e.g. the regulatory sequences.

Anchored PCR

In the basic PCR technique and the inverse PCR, one has to use two primers representing the sequences lying at the ends of sequences to be amplified. But sometimes, we may have knowledge about sequence at only one of the two ends of the DNA sequence to be amplified.

In such cases, anchored PCR may be used, which will utilize only one primer instead of two primers. In this technique, due to the use of one primer, only one strand will be copied first, after which a poly G tail will be attached at the end of the newly synthesized strand **(Fig. 4.2)**.

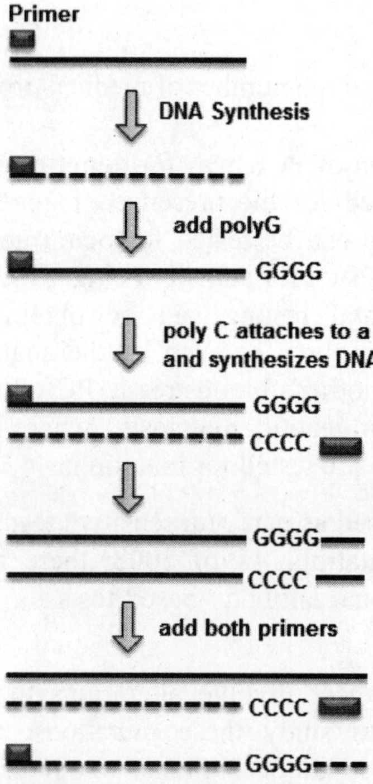

Fig. 4.2: Anchored PCR

This newly synthesized strand with poly G tail at its 3' end will then become template for the daughter strand synthesis utilizing an anchor primer with which a Poly C sequence is linked to complement with poly G of the template. In the next cycle, both the original primer and anchored primer will be used for gene amplification.

APPLICATIONS OF POLYMERASE CHAIN REACTION

Although PCR was first developed only a decade and a half ago, the simplicity and the versatility of the technique have ensured that is among the most ubiquitous of molecular genetic methodologies, which a wide range of general applications. During the last few years, with the improvement of PCR protocols, and due to the availability of automatic thermal cyclers commercially, the application of PCR have increased manifold. However for application of PCR, we need a pair of primers which could be based on the knowledge of nucleotide sequence of the DNA to be amplified. Therefore, non-availability of this information about the DNA segment or gene to be amplified becomes a limitation in the application of PCR, although in the study of DNA polymorphisms, this difficulty has been overcome through the use of random DNA primers.

Medical Applications

PCR has been applied to a large number of medical procedures:

- The first application of PCR was for **genetic testing**, where a sample of DNA is analyzed for the presence of genetic disease mutations. Prospective parents can be tested for being genetic carriers, or their children might be tested for actually being affected by a disease. DNA samples for prenatal testing can be obtained by amniocentesis, chorionic villus sampling, or even by the analysis of rare fetal cells circulating in the mother's bloodstream. PCR analysis is also essential to Preimplantation genetic diagnosis, where individual cells of a developing embryo are tested for mutations.

- PCR can also be used as part of a sensitive test for **tissue typing**, vital to organ transplantation. As of 2008, there is even a proposal to replace the traditional antibody-based tests for blood type with PCR-based tests.

- Many forms of cancer involve alterations to **oncogenes**. By using PCR-based tests to study these mutations, therapy regimens can sometimes be individually customized to a patient.

Infectious disease applications

Characterization and detection of infectious disease organisms have been revolutionized by PCR:

- The **Human Immunodeficiency Virus** (or **HIV**), responsible for **AIDS**, is a difficult target to find and eradicate. The earliest tests for infection relied on the presence of antibodies to the virus circulating in the bloodstream. However, antibodies don't appear until many weeks after infection, maternal antibodies mask the infection of a newborn, and therapeutic agents to fight the infection don't affect the antibodies. PCR tests have been developed that can detect as little as one viral genome among the DNA of over 50,000 host cells. Infections can be detected earlier, donated blood can be screened directly for the virus, newborns can be immediately tested for infection, and the effects of antiviral treatments can be quantified.

- Some disease organisms, such as that for **Tuberculosis**, are difficult to sample from patients and slow to be grown in the laboratory. PCR-based tests have allowed detection of small numbers of disease organisms (both live and dead), in convenient samples. Detailed genetic analysis can also be used to detect antibiotic resistance, allowing immediate and effective therapy. The effects of therapy can also be immediately evaluated.

- The spread of a **disease organism** through populations of domestic or wild animals can be monitored by PCR testing. In many cases, the appearance of new virulent sub-types can be detected and monitored. The sub-types of an organism that were responsible for earlier epidemics can also be determined by PCR analysis.

Forensic Applications

The development of PCR-based genetic (or DNA) fingerprinting protocols has seen widespread application in forensics:

- In its most discriminating form, **Genetic fingerprinting** can uniquely discriminate any one person from the entire population of the world. Minute samples of DNA can be isolated from a crime scene, and compared to that from suspects, or from a DNA database of earlier evidence or convicts. Simpler versions of these tests are often used to rapidly rule out suspects during a criminal investigation. Evidence from decades-old crimes can be tested, confirming or exonerating the people originally convicted.

- Less discriminating forms of DNA fingerprinting can help in **Parental testing**, where an individual is matched with their close relatives. DNA from unidentified human remains can be tested, and compared with that from possible parents, siblings, or children. Similar testing can be used to confirm the biological parents of an adopted (or kidnapped) child. The actual biological father of a newborn can also be confirmed (or ruled out).

Research Applications

PCR has been applied to many areas of research in molecular genetics:

- PCR allows rapid production of short pieces of DNA, even when nothing more than the sequence of the two primers is known. This ability of PCR augments many methods, such as generating hybridization probes for Southern or northern blot hybridization. PCR supplies these techniques with large amounts of pure DNA, sometimes as a single strand, enabling analysis even from very small amounts of starting material.

- The task of DNA sequencing can also be assisted by PCR. Known segments of DNA can easily be produced from a patient with a genetic disease mutation. Modifications to the amplification technique can extract segments from a completely unknown genome, or can generate just a single strand of an area of interest.

- PCR has numerous applications to the more traditional process of DNA cloning. It can extract segments for insertion into a vector from a larger genome, which may be only available in small quantities. Using a single set of 'vector primers', it can also analyze or extract fragments that have already been inserted into vectors. Some alterations to the PCR protocol can generate mutations (general or site-directed) of an inserted fragment.

- Sequence-tagged sites is a process where PCR is used as an indicator that a particular segment of a genome is present in a particular clone. The Human Genome Project found this application vital to mapping the cosmid clones they were sequencing, and to coordinating the results from different laboratories.

- An exciting application of PCR is the phylogenic analysis of DNA from ancient sources, such as that found in the recovered bones of Neanderthals, or from frozen tissues of Mammoths. In some cases the highly degraded DNA from these sources might be reassembled during the early stages of amplification.

- A common application of PCR is the study of patterns of gene expression. Tissues (or even individual cells) can be analyzed at different stages to see which genes have become active, or which have been switched off. This application can also use Q-PCR to quantitate the actual levels of expression

- The ability of PCR to simultaneously amplify several loci from individual sperm has greatly enhanced the more traditional task of genetic mapping by studying chromosomal crossovers after meiosis. Rare crossover events between very close loci have been directly observed by analyzing thousands of individual sperms. Similarly, unusual deletions, insertions, translocations, or inversions can be analyzed, all without having to wait (or pay for) the long and laborious processes of fertilization, embryogenesis, etc.

5

Cloning Vectors

A clone is an exact copy of an organism, organ, single cell, organelle or macromolecule. Cell lines for medical research are derived from a single cell allowed to replicate millions of times, producing masses of identical clones.

Gene cloning is the act of making copies of a single gene. Amplified genes are useful in many areas of research and for medical applications such as gene therapy. Selective amplification of genes depends on our ability to perform the following essential procedures.

CLONING VECTORS

Cloning vectors are small DNA vehicles designed to transport a foreign DNA fragment. DNA molecules are cloned for a variety of purposes including isolating novel genes, safeguarding DNA samples, facilitating sequencing, generating probes, and expressing recombinant protein in one or more host organisms. The DNA to be cloned can be produced from a number of sources including PCR, restriction digesting an existing vector, cDNA synthesis, RT-PCR, or genomic DNA preparations.

Few of the cloning vectors which are used in recombinant DNA technology include plasmids, bacteriophages, cosmids, phagemids and viruses. An ideal cloning vector should have the following characteristics:

1) It should have its own replicon and thereby be capable of autonomous replication in the host cell

2) It should carry one or more selectable marker function, in order to permit the recognition of cells carrying the parental form of the vector or a recombinant between the vector and foreign DNA sequences.

3) It should have a cloning site, a region containing unique restriction enzyme cleavage site into which a foreign DNA can be inserted without interference in plasmid's ability to replicate or to confer suitable phenotype.

Plasmids

Plasmids are circular, double-stranded DNA molecules that exist in bacteria and in the nuclei of some eukaryotic cells. They can replicate independently of the host cell. The size of plasmids ranges from a few kb to near 100 kb. Plasmids are autonomous elements, whose genomes exist in the cell as extra chromosomal units. They are self replicating circular duplex DNA molecules which are maintained in a characteristic number of copies in a bacterial cell, yeast cell or in eukaryotic cells. The cloning site is usually located in the middle of a selectable marker. The insertion of the cloned insert alters the associated phenotype. The circular plasmid DNA which is used as a vector can be cleaved at one site with the help of a restriction enzyme to give a linear DNA molecule.

A **foreign DNA** segment can now be inserted by joining the ends of a linearized plasmid DNA to the two ends of a foreign DNA, thus regenerating a recombinant DNA molecule that can now be separated by gel electrophoresis on the basis of its size.

Selection of recombinant DNA is facilitated by the resistance genes, which the plasmid may carry against one or more antibiotics. If a plasmid has two such genes conferring resistance against two antibiotics and if a foreign DNA insertion site lies within one of these two genes, then the chimeric vector loses resistance against one antibiotic. In such a situation, the parent vector in bacterial cells can be selected by resistance against two antibiotics and the recombinant DNA can be selected by retention of resistance against only one of the two antibiotics.

One of the standard cloning plasmid vectors widely used in the gene cloning experiments is pBR322 (derived from E. coli plasmid ColEl), which is a 4362 bp DNA and was derived by several alterations of the earlier cloning vector. pBR 322 has been named after Bolivar and Rodriguez, who prepared this vector. It has genes for resistance against two antibiotics (tetracycline and ampicillin), an origin of replication and a variety of restriction sites for cloning of restriction fragments obtained through cleavage with a specific restriction enzyme.

Another series of plasmids that are used as cloning vectors belong to pUC series (named after the place of their initial preparation, i.e., University of California). These plasmids are 2700 bp long and possess (Fig. 5.1). It has the following features:

(i) An ampicillin resistance gene.

(ii) An origin of replication derived from pBR 322 and

(iii) The lacZ gene derived from E. coli. Within the lac region is also found a polylinker sequence having unique restriction sites (identical to those found in phage M 13).

When DNA fragments are cloned in this region of pUC, the lac gene is inactivated. These plasmids, when transferred into an appropriate E. coli strain, which is lac (e.g. JMI03, JMI09), and grown in the presence of isopropyl thiogalactoside (IPTG, which behaves like lactose, and induces the synthesis of β-galactosidase enzyme) and X-gal (substrate for the enzyme), will give rise to white or clear colonies. On the other hand, pUC having no inserts and transferred into bacteria will have an active lacZ gene and therefore will produce blue colonies, thus permitting identification of colonies having pUC vector with DNA segment.

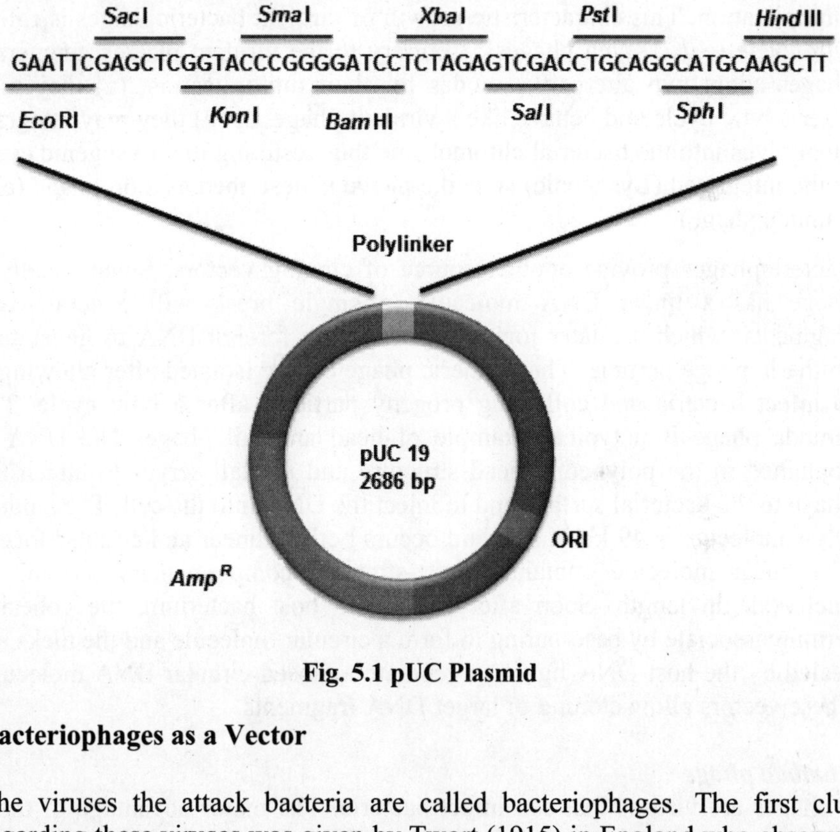

Fig. 5.1 pUC Plasmid

Bacteriophages as a Vector

The viruses the attack bacteria are called bacteriophages. The first clue regarding these viruses was given by Twort (1915) in England who observed the bacterial were and the lytic effect spread from one colony to the other. He speculated that the lysis of bacterial colonies would have been due to a virus. De Herelle (1917) repeated Twort's experiments at Pasteur Institute (Paris), rediscovered Tworts findings, named the phenomenon as Twort-de Herelle-phenomenon and coined the term bacteriophage (i.e., bacteria eater). Since then numerous bacteriophages have been discovered and they are the most extensively studied viruses.

The bacteriophages are commonly called phages. The phages possess dsDNA, ssDNA, dsRNA or ssRNA as genetic material. Three common forms (viz., tailed, cubic, and filamentous) of bacteriophages are known. The tailed bacteriophages from the largest group and are much studied; they have been named T-phages particularly T-even phages (T2, T4 and so on) and T-odd phages (TI, T3 and so on). However, on the basis of their interaction with the host bacterium, the phages are divided into two groups: virulent and temperate.

Virulent phages are those that normally lyses (destroy) the attacked host/bacterial cells; there is probably no alternative strategy for their multiplication. This characteristic growth of virulent bacteriophages is called lytic cycle (e.g. T-even phages). Contrary to the virulent phages, temperate phages adopt two alternative modes for their multiplication, (a) they may enter a lytic cycle and behave like a virulent phage, or (b) they may integrate themselves into the bacterial chromosome thus resulting in a Lysogenic cycle in the integrated (Lysogenic) state the phage is described as a prophage (e.g. Lambda phage).

Bacteriophages provide another source of cloning vectors. Since usually a phage has a linear DNA molecule, a single break will generate two fragments, which are later joined together with foreign DNA to generate a chimeric phage particle. The chimeric phage can be isolated after allowing it to infect bacteria and collecting progeny particles after a lytic cycle. The lambda phage is a typical example of head and tail phage. The DNA is contained in the polyhedral head structure and the tail serves to attach the phage to the bacterial surface and to inject the DNA into the cell. The lambda DNA molecule is 49 kb in size and occurs both in linear and circular forms. The linear molecule contains single-stranded complementary termini 12 nucleotide in length. Soon after entering a host bacterium, the cohesive termini associate by base-paring to form a circular molecule and the nicks are sealed by the host DNA ligase to generate a closed circular DNA molecule. These vectors allow cloning of larger DNA fragments.

Lambda phage

λ phages are viruses that can infect bacteria. The major advantage of the λ phage vector is its high transformation efficiency, about 1000 times more efficient than the plasmid vector.

The DNA to be cloned is first inserted into the λ DNA, replacing a nonessential region. Then, by an **in vitro assembly system**, the λ virion carrying the recombinant DNA can be formed. The λ genome is 49 kb in length which can carry up to 25 kb foreign DNA.

The extreme ends of the λ DNA are known as **COS sites**, each are single stranded, 12 nucleotides long. Because their sequences are complementary to each other, one end of λ DNA may base-pair with the other end of a different λ DNA, forming concatemers. The two ends of a λ DNA may also bind together, forming a circular DNA. In the host cell, the λ DNA circularizes because ligase may seal the join of the COS sites.

In the assembly process of λ virions, two proteins Nu1 and A can recognize the COS site, directing the insertion of the λ DNA between them into an empty head. The filled head is then attached to the tail, forming a complete λ virion. The whole process normally takes place in the host cell. However, to prepare the λ virion carrying recombinant λ DNA, the following **in vitro assembly system** is commonly used (Fig. 5.2).

Proteins Nu1 and A are encoded by the genes in the λ genome. If the two genes are mutated, λ DNA cannot be packaged into the pre-assembled head. Because tails attach only to filled heads, the cell will accumulate separate empty heads and tails, which can then be extracted. When the extract is mixed with recombinant λ DNA and proteins Nu1 and A, the complete λ virion carrying recombinant λ DNA will be assembled.

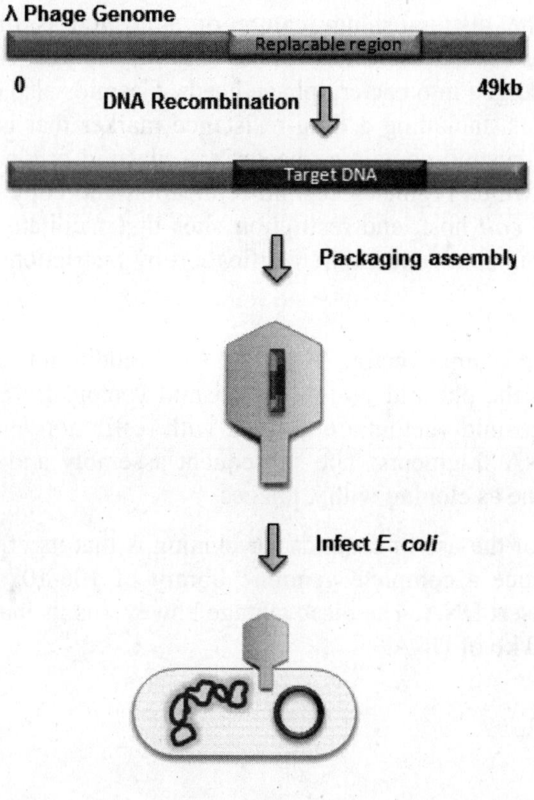

Fig. 5.2 Schematic drawing of the DNA cloning using λ phages as vectors

M13 phage

M 13 is a filamentous bacteriophage of E. coli and contains a 7.2 kb long single-stranded circular DNA (Fig. 5.3). M13 phage has been variously modified to give rise to M13 mp series of cloning vectors which can be used for cloning of a wide variety of DNA fragments and have the unique advantage of generating large quantities of DNA molecules that carry the sequence of one strand of the foreign DNA. Such single-stranded DNAs are the templates of choice for:

1) DNA sequencing by the dideoxy chain termination method.
2) Generating DNA probes for hybridization that are radio labeled in only one strand (Fig. 5.5) and
3) Site-directed mutagenesis using synthetic oligonucleotide (Fig. 5.4).

Cosmids

Cosmids are cloning vectors that were developed to enable large fragments of DNA to be cloned and maintained. Cosmid vectors allow the cloning of fragments up to 45 kilobases (kb) and are commonly used in genomic library construction. The distinguishing feature of a cosmid is the presence of bacteriophage λ *cos* sites, which enable the vector, and cloned genomic DNA, to be packaged into bacteriophage heads. Cosmids also contain several other components, including a drug-resistance marker that is necessary for the selection of cosmid-containing bacteria, a plasmid origin of replication, usually ColE1, which regulates cosmid replication and copy number within the *Escherichia coli* host, and restriction sites that facilitate cloning of the desired DNA fragments and insert verification by restriction digestion (Fig. 5.6).

Cloning by using cosmid vectors (Fig. 5.7). **(a)** In addition to *amp*^{r}, ORI, and polylinker as in the plasmid vector, the cosmid vector also contains a COS site. **(b)** After cosmid vectors are cleaved with restriction enzyme, they are ligated with DNA fragments. The subsequent assembly and transformation steps are the same as cloning with λ phages.

The advantage of the use of cosmids for cloning is that its efficiency is high enough to produce a complete genomic library of 106-107 clones from a mere 1 mg of insert DNA. The disadvantage however is its inability to accept more than 40-50 kb of DNA.

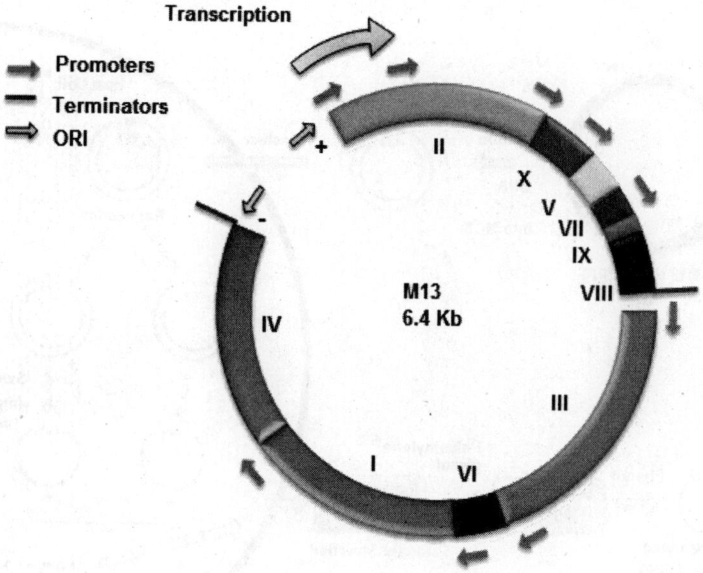

Fig. 5.3 The M13 Vector

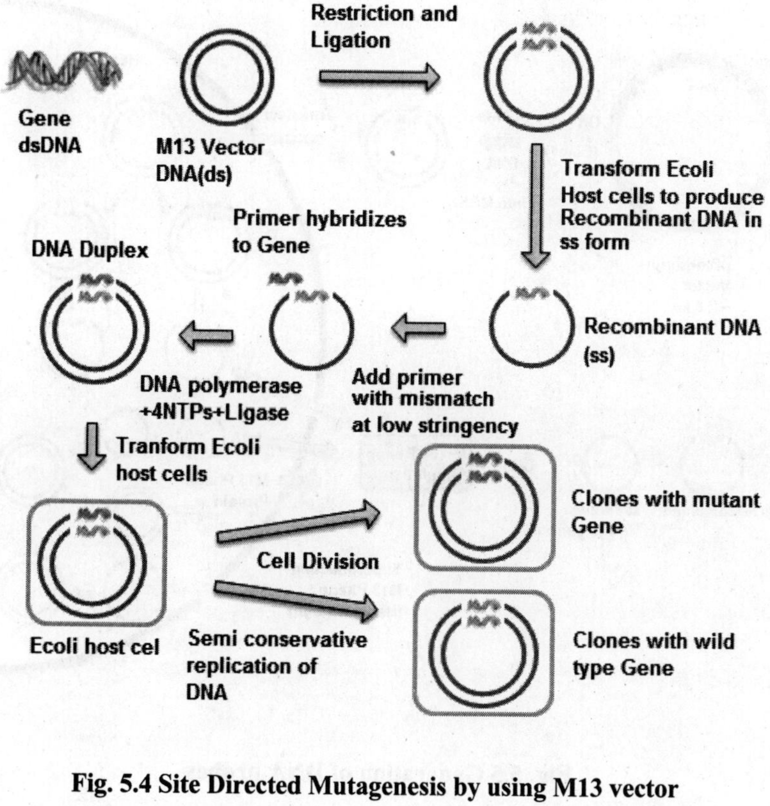

Fig. 5.4 Site Directed Mutagenesis by using M13 vector

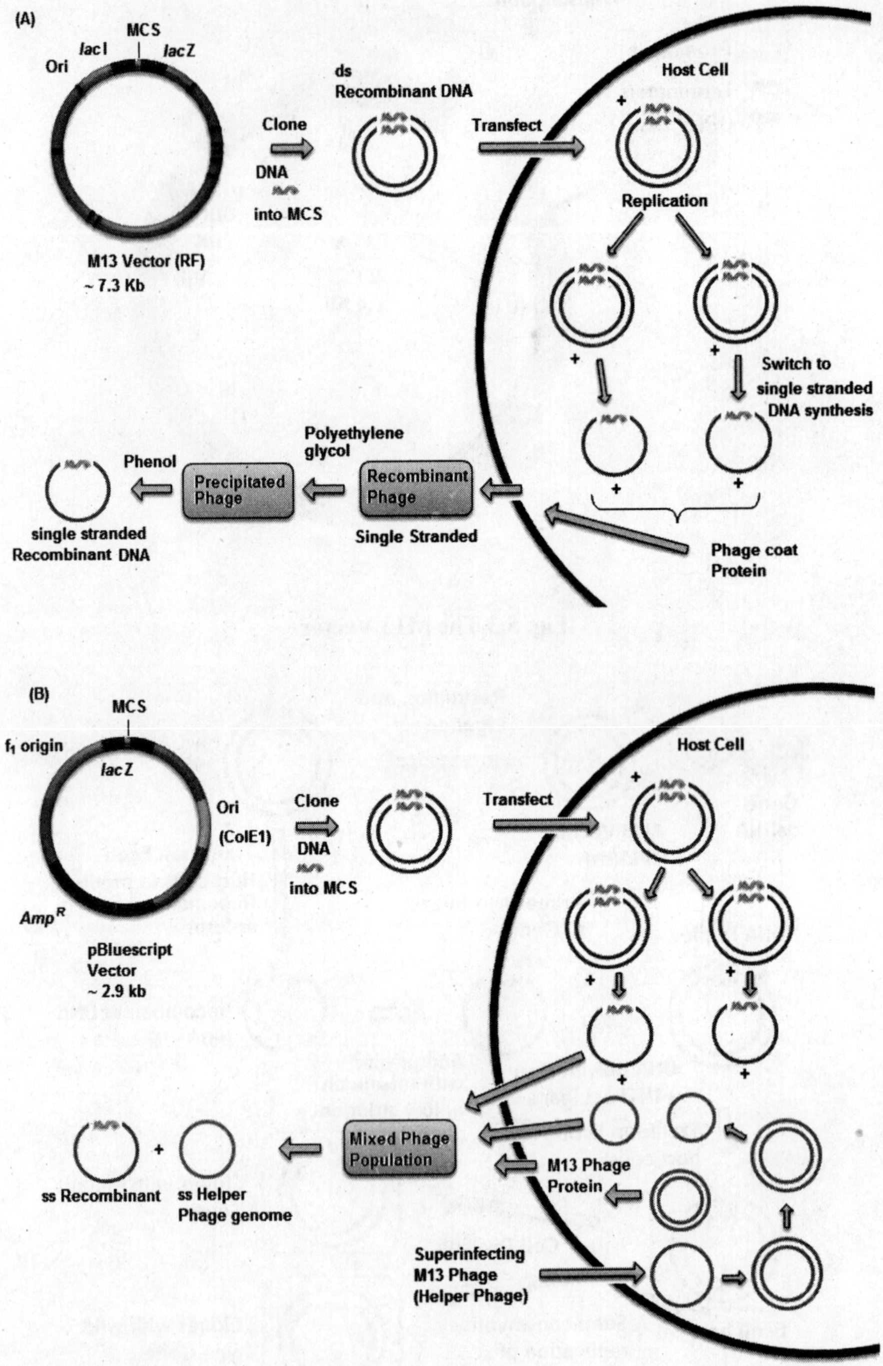

Fig. 5.5 Generation of DNA probes

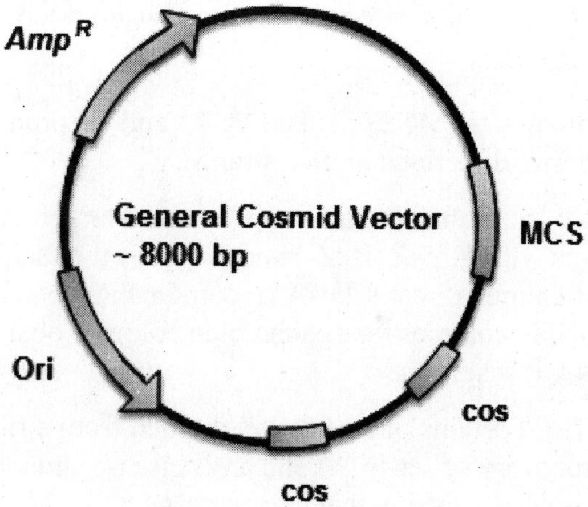

Fig. 5.6 Representation of a general cosmid vector

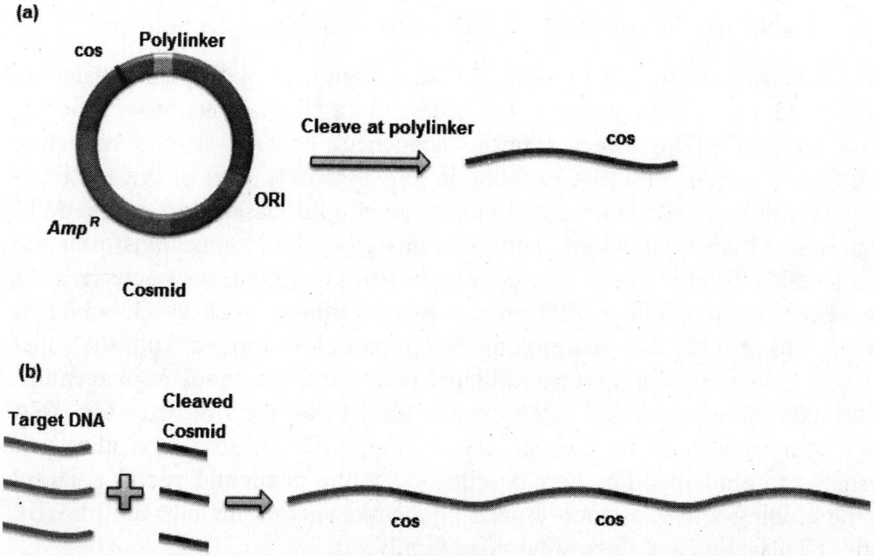

Fig. 5.7 Construction of cosmid vector

Phagemids

Phagemids are used as vectors and are prepared artificially by combining features of phages with plasmids, as the name suggests. One such phagemid, which is commonly used in molecular biology laboratories, is pBluescript II KS, which is derived from pUC 19 and is 2961 bp long. The KS designation indicates the orientation of polylinker, such that the transcription of lacZ gene proceeds from restriction site for Kpn I to that for Sac I. This phagemid has the following features:

1) A multi cloning site (MCS) flanked by T3 and T7 promoters to be read in opposite directions on two strands.
2) An inducible lac promoter (lac I), upstream of lacZ region, which complements with *E. coli* (lacZ-) and provides the facility for selection of chimeric vector DNA (recombinant vector) using the criteria of white colonies (as against blue colonies obtained if no foreign DNA is inserted).
3) fl (+) and fl (-) origins of replication derived from a filamentous phage for recovery of sense (+) and antisense (-) strands of lacZ gene, when host is co-infected with a helper phage.
4) An origin of replication (ColE1 ori) derived from plasmid, and used in the absence of helper phage.
5) A gene having ampicillin resistance for antibiotic selection.

Phagemid vectors are plasmids which have been artificially manipulated so as to contain a small segment of the genome of a filamentous phage, such as M13, fd or fl. The selected phage sequences contain all the *cis*-acting elements required for DNA replication and assembly into phage particles. They permit successful cloning of inserts several kilobases long (unlike M13 vectors in which such inserts tend to be unstable). Following transformation of a suitable *E. coli* strain with a recombinant phagemid, the bacterial cells are *super infected* with a filamentous **helper phage**, such as fl, which is required to provide the coat protein. Phage particles secreted from the super infected cells will be a mixture of helper phage and recombinant phagemids. The mixed single-stranded DNA population can be used directly for DNA sequencing because the primer for initiating DNA strand synthesis is designed to bind specifically to a sequence of the phagemid vector adjacent to the cloning site. Commonly used phagemid vectors include the pEMBL series of plasmids and the p Bluescript family.

Artificial Chromosomes

pYAC or Yeast artificial chromosome

This carries centromere and telomere sequences and therefore can be used to obtain artificial chromosomes. One example is pYAC3, which is essentially a pBR 322 into which a number of yeast genes have been inserted. Two of these genes, URA3 and TRP 1, are seen as selectable markers for Yip5 and YRp7 respectively. The DNA fragment that carries TRP 1 gene also contains an origin of replication, but in pYAC3, this fragment is extended further to include the sequence called CEN4, which is the DNA from the centromere region of chromosome 4.

The other component, the telomeres are provided by two sequences called TEL. This plasmid also contains SUP4, which is a selectable marker into which new DNA is inserted during the cloning experiment. For cloning, the vector is first restricted with a combination of Bam HI and Sna BI, cutting the molecule into three fragments.

The Barn HI fragment is removed, leaving two arms, each bounded by one TEL sequence and one Sna BI site. The DNA to be cloned, which must have blunt ends (Sna BI, a blunt end cutter, recognizes TACGTA) is ligated between the two arms producing the artificial chromosome.

Procedure

1) The target DNA is partially digested by EcoRI and the YAC vector is cleaved by EcoRI and BamHI (Fig. 5.8).
2) Ligate the cleaved vector segments with a digested DNA fragment to form an artificial chromosome.
3) Transform yeast cells to make a large number of copies.

YACs have two disadvantages:

(i) Cloning efficiency is low (1000 clones/μg DNA as against 10^6 - 10^7 clones/μg DNA for cosmids), thus making it impractical to generate complete genomic libraries through the use of YACs.

(ii) It is not possible to recover large amount of pure insert DNA from individual clones; selective amplification of YACs DNA has recently allowed this problem to be overcome. Both these problems have been overcome in BACs and PACs.

YACs as vectors for cloning large DNA segments (100-300 kb)
Genome analysis and map based cloning in higher plants and animals required isolation, of large pieces of DNA, which cannot be cloned in plasmid or phage vectors. A variety of vectors are now available; which are suitable for cloning such large DNA fragments suitable for preparation of high-resolution physical maps of higher eukaryotes (both animals and plants).

These vectors, 0.5 to 2 megabase (Mb) in length, particularly include yeast artificial chromosomes, (YACs) capable of accepting fragments of 100-2000 kbp and PI vectors capable of cloning fragments of upto 100 kbp (kilo base pairs).

Mammalian artificial chromosome (MAC) vectors have also been designed, which' are capable of cloning fragments of more than 1000 kilo base pairs. Plasmids based on F-factors were also designed which could accept fragments upto 120kbp in length (for YAC and MAC vector, see later in this chapter). Of these vectors, YACs are most commonly used, but following difficulties have been encountered while using YACs:

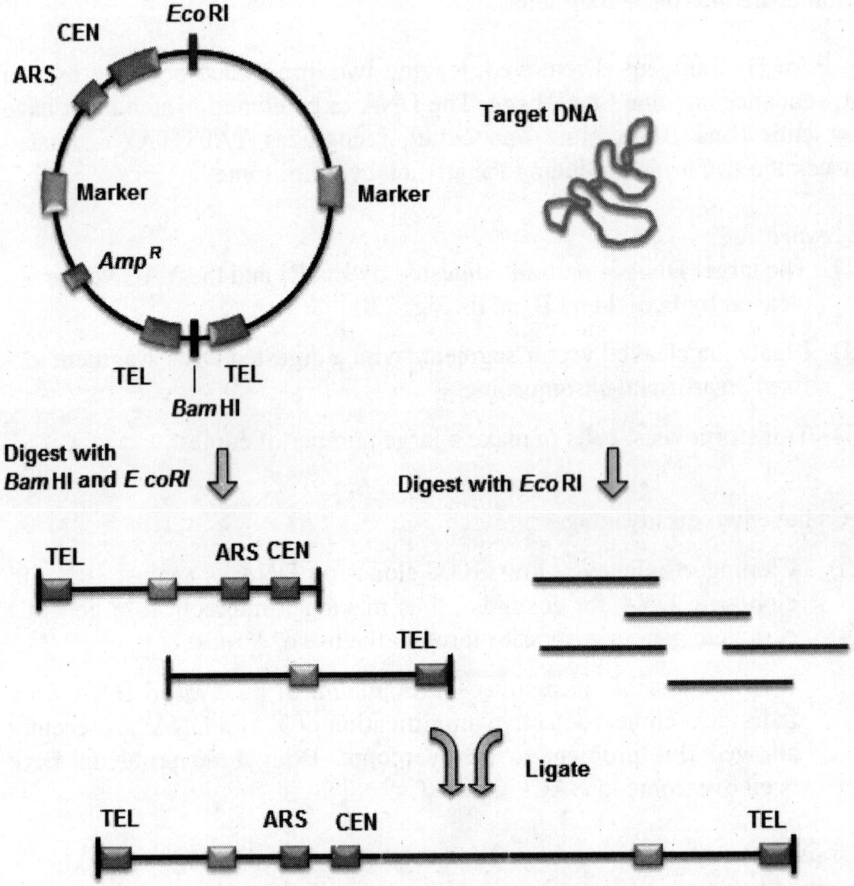

Fig. 5.8 Cloning by the yeast artificial chromosome (YAC) vector.

(i) In YAC libraries, a fraction of clones result due to co-cloning events giving single clones with non-contiguous fragments-these are described as chimeric vectors.

(ii) YAC clones also exhibit some degree of instability due to deletions/rearrangements in the cloned inserts.

(iii) YACs are similar to yeast normal chromosomes in size thus making it difficult to separate them by simple methods.

Bacterial artificial chromosomes (BACs)

In order to overcome the above difficulties associated with YAC vector, a bacterial cloning system based on E. coli F factor was designed which was capable of cloning fragments of upto 300-350kb. These were described as **bacterial artificial chromosomes** (BACs) and is 'user friendly' being a bacterial system. BAC vectors are superior to other bacterial systems, based on high to medium copy number of replicons, since they show structural instability of inserts, deleting or rearranging portions of cloned DNA (Fig. 5.9).

However, the F factor has regulatory genes that regulate its own replication and controls its copy number. These regulatory genes include (i) oriS and repE which mediate unidirectional replication and (ii) parA and parR, which maintain the copy number to 1 or 2 per E. coli genome. These essential genes of F factor are incorporated in every BAC vector (pBAC), which also has a chloroamphenicol resistance gene as a marker and a cloning segment.

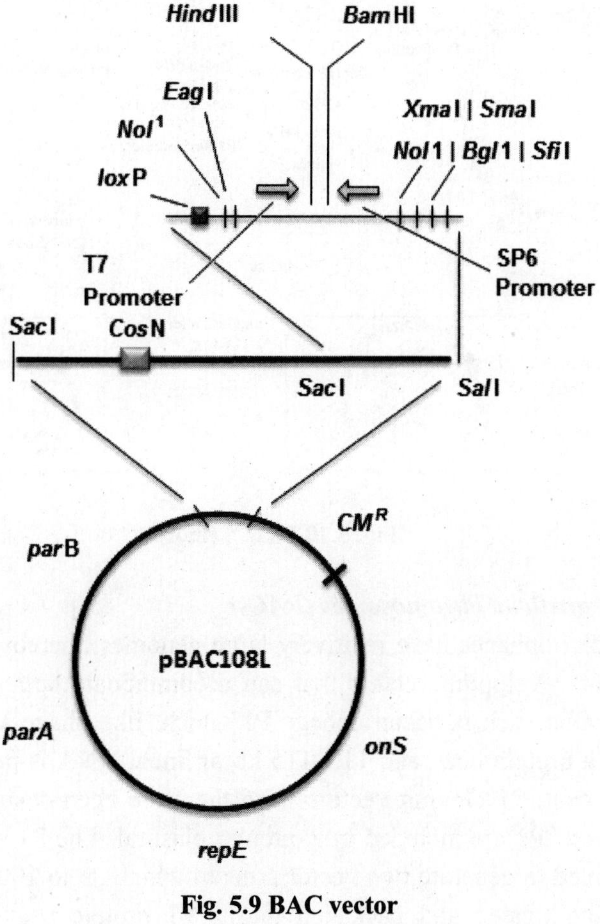

Fig. 5.9 BAC vector

Cloning segment includes the following sequences:

(i) Phage lambda cosN site (providing a fixed position for specific cleavage with lambda terminase) and loxP site (providing a position for cleavage due to PI Cre protein in presence of loxP oligonucleotide); these two sites allow generation of ends that can be used for restriction site mapping to arrange the clones in an ordered array.

(ii) Two cloning sites (HindIII and BamHI) and

(iii) Several C + G rich restriction sites (NotI, EagI, XmaI, SmaI, Bg I1 and SfiI) for potential excision of inserts; the cloning site is flanked by T7 and SP6 promoters for generating RNA probes for chromosome walking and for DNA sequencing. BAC libraries have already been prepared in humans, mouse, rice, wheat, Lotus, etc.

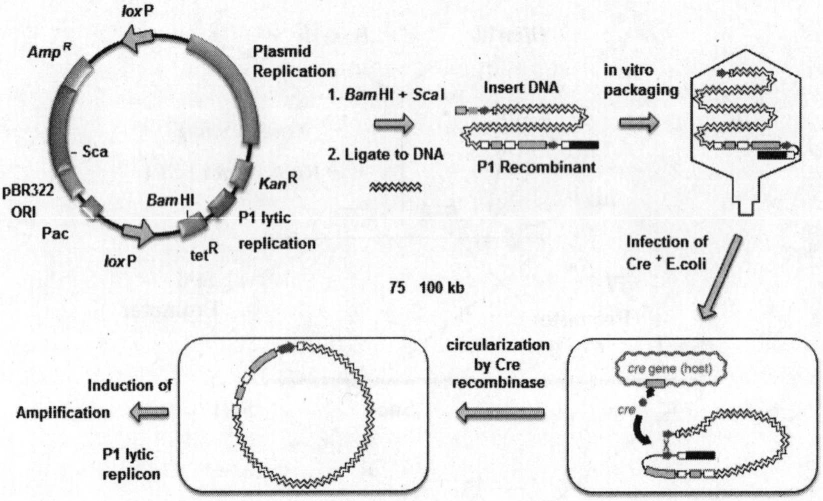

Fig. 5.10 PAC Vector

P1-derived artificial chromosomes (PACs)

Certain bacteriophages have relatively large genomes, thereby affording the potential for developing vectors that can accommodate large foreign DNA fragments. One such is bacteriophage P1 which, like phage λ, packages its genome in a protein coat, and 110–115 kb of linear DNA is packaged in the P1 protein coat. P1 cloning vectors have therefore been designed in which components of P1 are included in a circular plasmid. The P1 plasmid vector can be cleaved to generate two vector arms to which up to 100 kb of foreign DNA can be ligated and packaged into a P1 protein coat *in vitro*. The recombinant P1 phage can be allowed to adsorb to a suitable host, following which the recombinant P1 DNA is injected into the cell, circularizes and can be amplified (Fig. 5.10).

An improvement on the size range of inserts accepted by the basic P1 cloning system has been the use of bacteriophage T4 *in vitro* packaging systems with P1 vectors which enables the recovery of inserts up to 122 kb in size. More recently, features of the P1 and F-factor systems have been combined to produce P1-derived artificial chromosome (PAC) cloning system.

Mammalian artificial chromosomes (MACs)

Mammalian artificial chromosome (MAC) vectors are designed to be able to replicate, segregate and express in a mammalian cell like any other mammalian chromosome along with other chromosomes. Since it will be an independent chromosome, with all the functional elements (telomeres, origins of replication, centromere, etc.), MAC will not be integrated with the genome and can be used as a vector maintaining a single copy per cell.

It could carry large fragments of DNA representing an intact eukaryotic split gene with exons and introns permitting its normal expression regulated by the associated promoter sequences. In view of this, MACs are considered to be suitable for gene therapy, where the inserted DNA will be fully expressed, yet stably maintained without affecting the host genome.

During 1994-96 the possibility of the construction of MACs was still being explored. Firstly, only telomere sequences were available to be used in MACs. The other two components, i.e. the replication origins and the centromere sequences were not available to be cloned and used for the construction of MACs.

Sequences for attachment to nuclear matrix or matrix attachment regions (MAR) may also be needed. The elements of YACs are constructed in E. coli, ligated in vitro and then transferred to yeast cells, but for MACs this may be difficult, since cloning of large mammalian DNA segments in E. coli and yeast may be difficult.

Therefore, one may design a MAC of the smallest possible size, then construct it in yeast and transfer it to mammalian chromosome. These problems associated with the construction of MACs and their insertion into mammalian cells was partly overcome later.

6

Gene Sequencing

OVERVIEW

In genetics terminology, DNA sequencing is the process of determining the nucleotide order of a given DNA fragment. Currently, almost all DNA sequencing is performed using the chain termination method, developed by Frederick Sanger. This technique uses sequence-specific termination of an *in vitro* DNA synthesis reaction using modified nucleotide substrates.

WHY SEQUENCE DNA?

The sequence of DNA encodes the necessary information for living things to survive and reproduce. Determining the sequence is therefore useful in 'pure' research into why and how organisms live, as well as in applied subjects. Because of the key nature of DNA to living things, knowledge of DNA sequence may come in useful in practically any biological research. For example, in medicine it can be used to identify, diagnose and potentially develop treatments for genetic diseases. Similarly, research into pathogens may lead to treatments for contagious diseases. Biotechnology is a burgeoning discipline, with the potential for many useful products and services.

SANGER SEQUENCING

In chain terminator sequencing (Sanger sequencing), extension is initiated at a specific site on the template DNA by using a short oligonucleotide 'primer' complementary to the template at that region. The oligonucleotide primer is extended using a DNA polymerase, an enzyme that replicates DNA. Included with the primer and DNA polymerase are the four deoxynucleotide bases (DNA building blocks), along with a low concentration of a chain terminating nucleotide (most commonly a **di**-deoxynucleotide). Limited incorporation of the chain terminating nucleotide by the DNA polymerase results in a series of related DNA fragments that are terminated only at positions where that particular nucleotide is used. The fragments are then size-separated by electrophoresis in a slab polyacrylamide gel, or more commonly now, in a narrow glass tube (capillary) filled with a viscous polymer.

There are two sub-types of chain-termination sequencing. In the original method, the nucleotide order of a particular DNA template can be inferred by performing four parallel extension reactions using one of the four chain-terminating bases in each reaction. The DNA fragments are detected by labeling the primer with radioactive phosphorous prior to performing the sequencing reaction. The four reactions would then be run out in four adjacent lanes on a slab polyacrylamide gel.

A development of this method used four different fluorescent dye-labeled primers. This has the advantage of avoiding the need for radioactivity; increasing safety and speed, and also that the four reactions can be combined and run in a single Gel lane, if they can be distinguished. This approach is known as 'dye primer sequencing'.

An alternative to the labeling the primer is to label the terminators instead, commonly called 'dye terminator sequencing'. The major advantage of this approach is the complete sequencing set can be performed in a single reaction, rather than the four needed with the labeled-primer approach. This is accomplished by labeling each of the dideoxynucleotides chain-terminators with a separate fluorescent dye, which fluoresces at a different wavelength. This method is easier and quicker than the dye primer approach, but may produce more uneven data peaks (different heights), due to a template dependent difference in the incorporation of the large dye chain-terminators. This problem has been significantly reduced with the introduction of new enzymes and dyes that minimize incorporation variability.

This method now used for the vast majority of sequencing reactions as it is both simpler and cheaper. The major reason for this is that the primers do not have to be separately labeled (which can be a significant expense for a single-use custom primer), although this less of a concern with frequently used 'universal' primers.

Modern automated DNA sequencing instruments are able to sequence as many as 384 fluorescently labeled samples in a batch (run) and perform as many as 24 runs a day. These perform only the size separation and peak reading; the actual sequencing reaction(s), cleanup and resuspension in a suitable buffer must be performed separately.

To produce detectable labeled products from the template DNA, 'cycle sequencing' is most commonly performed. This approach uses repeated (25 - 40) rounds of primer annealing, DNA polymerase extension and disassociation (melting) of the template DNA strands. The major advantages of cycle sequencing is the more efficient use of the expensive sequencing reagent (Big Dye) and the ability to sequence templates with strong secondary structures such as hairpins or GC-rich regions. The different stages of cycle sequencing are performed by altering the temperature of the reaction using a PCR thermal cycler. This relies on the fact that complementary DNA will anneal at lower temperatures and disassociate at higher temperatures. An important part of making this possible is the use of DNA polymerase from a thermophillic organism, which is not rapidly denatured at the high (>95C) temperatures involved.

MAXAM-GILBERT SEQUENCING

At around the same time that the Sanger sequencing method was introduced, Maxam and Gilbert developed a method of DNA sequencing based on chemical modification of DNA followed by its subsequent cleavage. This method was initially popular since purified DNA could be used directly, while the initial Sanger method required that each read start be cloned for production of single-stranded DNA. As the chain termination method has been developed and improved, Maxam-Gilbert sequencing has fallen out of favour due to its technical complexity, the need for use of hazardous chemicals, and difficulties with scale-up.

OTHER DNA SEQUENCING METHODS

Other sequencing techniques which are under development, and may offer benefits over the conventional methods, include:

a) Sequencing by Hybridization

b) Pyrosequencing

c) Nanopore sequencing

LARGE-SCALE SEQUENCING STRATEGIES

Current methods can directly sequence only short lengths of DNA at a time. For example, modern sequencing machines using the Sanger method can achieve a maximum of around 1000 base pairs. This limitation is due to the geometrically decreasing probability of chain termination at increasing lengths, as well as physical limitations on gel size and resolution.

It is often necessary to obtain the sequence of much larger regions. For example, even simple bacterial genomes contain millions of base pairs, and the human genome has more than 3 billion. Several strategies have been devised for large-scale DNA sequencing, including primer walking (see also chromosome walking) and shotgun sequencing. These involve taking many small *reads* of the DNA through the Sanger method and subsequently assembling them into a contiguous sequence. The different strategies have different tradeoffs in speed and accuracy; for example, the shotgun method is the most practical for sequencing large genomes, but its assembly process is complex and potentially error-prone.

It is easier to obtain high quality sequence data when the desired DNA is purified and amplified from any contaminants that may be in the original sample. This can be achieved through PCR if it is practical to design primers that cover the entire desired region. Alternatively, the sample can be cloned using a bacterial vector, harnessing bacteria to "grow" copies of the desired DNA a few thousand base pairs at a time. Most large-scale sequencing efforts involve the preparation of a large *library* of such clones.

SEQUENCING OF GENE OR GENE SEGMENT

Once a gene or a DNA fragment has been cloned, its further study involves DNA sequencing. The technique of DNA sequencing was very laborious, till 1975, when a breakthrough was made in DNA sequencing methods.

Two different methods are now routinely used for determination of DNA sequences.

Maxam and Gilbert's Chemical Degradation Method

In this method, following steps are involved (Fig. 6.1):

(i) Label the 3'ends of DNA with ^{32}P.

(ii) Separate two strands, both labeled at 3'ends.

(iii) Divide the mixture in four samples, each treated with a different reagent having the property of destroying either only G, or only C, or 'A and G' or 'T and C. The concentration of reagent is so adjusted that 50% of target base is destroyed, so that fragments of different sizes having 32p are produced.

(iv) Electrophoreses each of the four samples in four different lanes of the gel.

(v) Autoradiograph the gel and determine the sequence from positions of bands in four lanes.

Fred Sanger's Dideoxynucleotide Synthetic Method

Fred Sanger (who won Nobel Prize twice) had initially developed a method for DNA sequencing, which utilized DNA polymerase to extend DNA chain length. This was termed plus minus method. Subsequently, he developed a more powerful method, utilizing single stranded DNA as template for DNA synthesis, in which 2', 3' dideoxynucleotides were incorporated leading to termination of DNA synthesis.

These dideoxynucleotides are used as triphosphate and can be incorporated in a growing chain, but they terminate synthesis, since they can not form a phosphodiester bond with next incoming deoxynucleotide triphosphate (dNTP).

Following steps are involved in Sanger's dideoxy method for DNA sequencing (Fig. 6.2):

(i) Four reaction tubes are set up, each containing single stranded DNA sample (cloned in M13 phage) to be sequenced, all four dNTPs (radioactively labeled) and an enzyme for DNA synthesis (DNA polymerase I). Each tube also contains a small amount (much smaller amount relative to four dNTPs) of one of the four ddNTP, so that four tubes have each a different ddNTP, bringing about termination at a specific base adenine (A), cytosine (C), guanine (G) and thymine (T).

(ii) The fragments, generated by random incorporation of ddNTP leading to termination of reaction, are then separated by electrophoresis on a high resolution Polyacrylamide gel. This is done for all the four reaction mixtures on adjoining lanes in the gel.

(iii) The gel is used for autoradiography so that the position of different bands in each lane can be visualized.

(iv) The bands on autoradiogram can be used for getting the DNA sequence.

A variant of the above dideoxy method was later developed, which has allowed the production of automatic sequencers. In this new approach, of a different fluorescent dye is tagged to the oligonucleotide primer in each of the four reaction tubes (blue for: A, red for C, etc.). The four reaction mixtures are pooled and electrophoreses together in a single Polyacrylamide of the tube.

A high sensitivity fluorescence detector, placed near the bottom of the tube, measures the amount, of each fluorophore as a function of time. The sequence is determined from the temporal order of peaks corresponding to four different dyes.

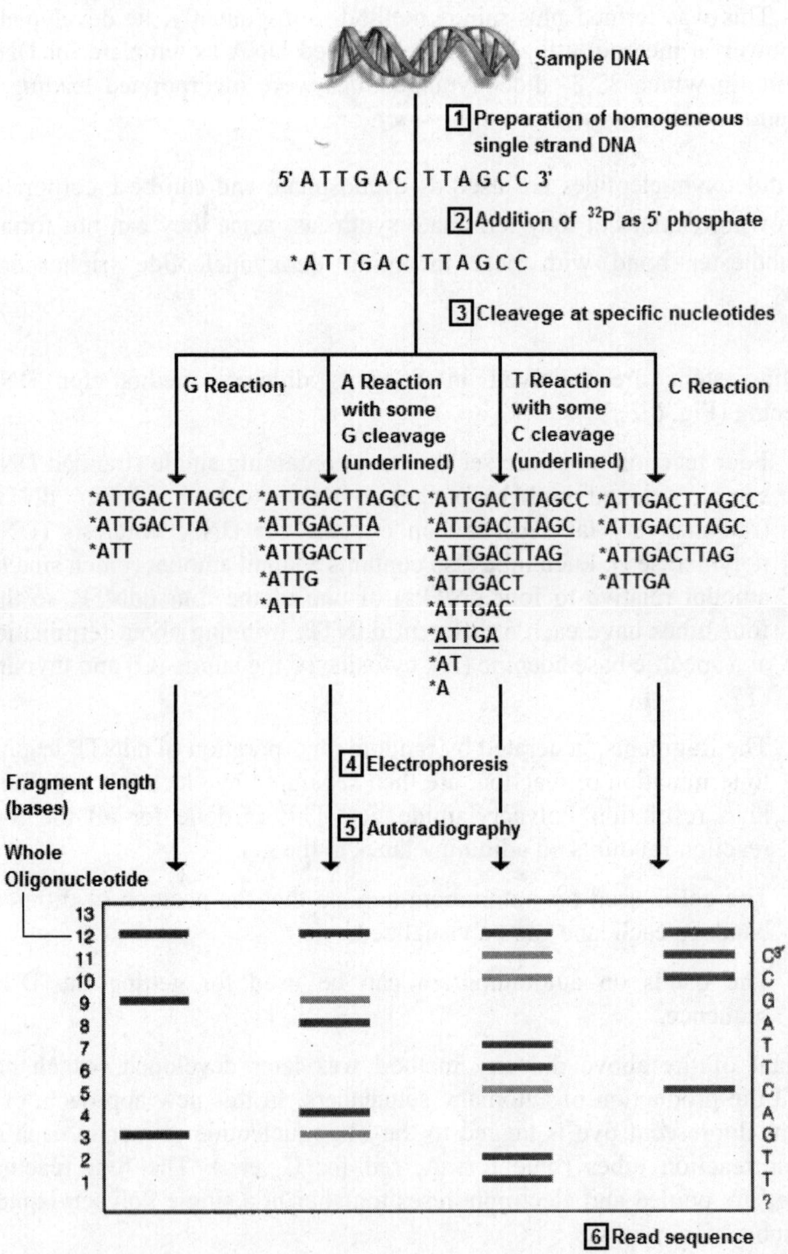

Fig. 6.1 Chemical degradation method

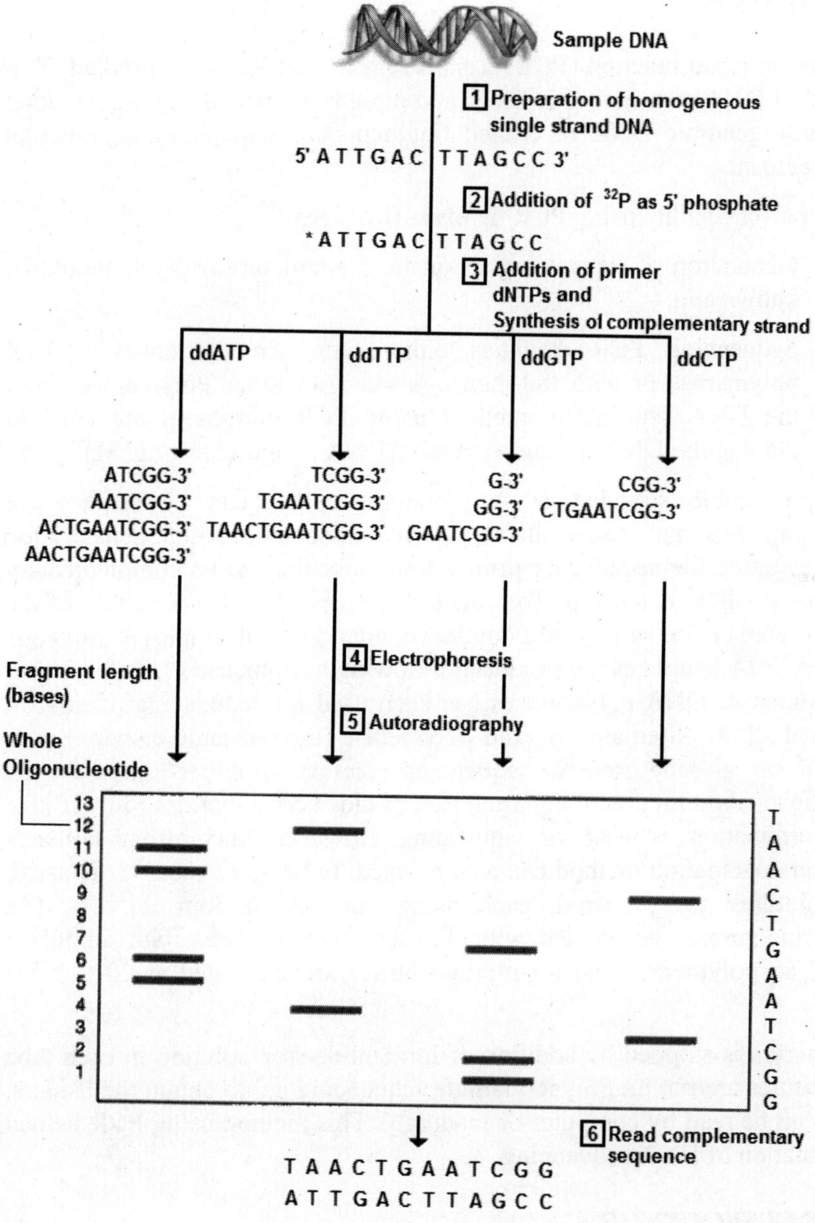

Fig. 6.2 Chain Termination Method

Direct DNA Sequencing Using PCR OR Ligation Mediated PCR - LMPCR

Polymerase chain reaction (PCR) has also been used for DNA product. This method of DNA sequencing is faster and more reliable and can utilize either the whole genomic DNA or cloned fragments for sequencing a particular DNA segment.

The DNA sequencing using PCR involves two steps:

(i) Generation of sequencing template, the structure of gene should be known and

(ii) Sequencing PCR products either with the thermolabile DNA polymerase or with the thermo stable Taq DNA Polymerase. Thus the DNA sequencing method using PCR eliminates the need of cloning the DNA in single stranded DNA phage vector i.e. M13.

Although double stranded DNA product of PCR can be utilized for sequencing, this may cause difficulty due reassociation after denaturation thus preventing the sequencing primer from annealing to its complementary sequence to allow extension. To reduce this problem, either a variant of the standard method for sequencing double stranded DNA is employed or single stranded DNA templates are produced following asymmetric PCR .A number of thermolabile DNA polymerases has been used for sequencing of in vitro amplified DNA. Alternatively, Taq polymerase (thermostable enzyme) used for PCR can also be used for sequencing reaction. In either case, Sanger's synthetic' method involving incorporation of dideoxynucleotides (ddNTP) for chain termination, is used for sequencing. However, Maxam and Gilbert's chemical degradation method can also be used. In Sanger's method, as usual, four mixtures are prepared, each using one of the four ddNTP. The sequencing primer is labelled with 32p and, the mixtures with amplified DNA, Taq polymerase and appropriate buffer are incubated at 70°C for 5 min.

The reaction is stopped by addition of formamide stop solution in each tube and mixtures are run on Polyacrylamide sequencing gel to obtain the ladders, which can be read by computer or manually. This method using PCR helped in automation of DNA sequencing.

GENOME SEQUENCING

A genome is all of a living thing's genetic material. It is the entire set of hereditary instructions for building, running, and maintaining an organism, and passing life on to the next generation.

In most living things, the genome is made of a chemical called DNA. The genome contains genes, which are packaged in chromosomes and affect specific characteristics of the organism. In short, the genome is divided into chromosomes, chromosomes contain genes, and genes are made of DNA.

Genome sequencing is figuring out the order of DNA nucleotides, or bases, in a genome—the order of As, Cs, Gs, and Ts that make up an organism's DNA. The human genome is made up of over 3 billion of these genetic letters.

Today, DNA sequencing on a large scale—the scale necessary for ambitious projects such as sequencing an entire genome—is mostly done by high-tech machines. Much as your eye scans a sequence of letters to read a sentence, these machines "read" a sequence of DNA bases.

A DNA sequence that has been translated from life's chemical alphabet into our alphabet of written letters might look like this:

ATCGGCCTAGGCTAGCTAGAATTCTGAAAGGTCT

By itself, not a whole lot. Genome sequencing is often compared to "decoding," but a sequence is still very much in code. In a sense, a genome sequence is simply a very long string of letters in a mysterious language. So sequencing the genome doesn't immediately lay open the genetic secrets of an entire species. Even with a rough draft of the human genome sequence in hand, much work remains to be done. Scientists still have to translate those strings of letters into an understanding of how the genome works: what the various genes that make up the genome do, how different genes are related, and how the various parts of the genome are coordinated. That is, they have to figure out what those letters of the genome sequence mean.

Significance

Sequencing the genome is an important step towards understanding it. At the very least, the genome sequence will represent a valuable shortcut, helping scientists find genes much more easily and quickly. A genome sequence does contain some clues about where genes are, even though scientists are just learning to interpret these clues.

Scientists also hope that being able to study the entire genome sequence will help them understand how the genome as a whole works—how genes work together to direct the growth, development and maintenance of an entire organism.

Finally, genes account for less than 25 percent of the DNA in the genome, and so knowing the entire genome sequence will help scientists study the parts of the genome outside the genes. This includes the regulatory regions that control how genes are turned on an off, as well as long stretches of "nonsense" or "junk" DNA—so called because we don't yet know what, if anything, it does.

SEQUENCING WHOLE GENOME

The whole genome can't be sequenced all at once because available methods of DNA sequencing can only handle short stretches of DNA at a time. So instead, scientists must break the genome into small pieces, sequence the pieces, and then reassemble them in the proper order to arrive at the sequence of the whole genome.

There are two approaches to the task of cutting up the genome and putting it back together again. One strategy, known as the **"clone-by-clone"** approach, involves first breaking the genome up into relatively large chunks, called clones, about 150,000 base pairs (bp) long. Scientists use genome mapping techniques to figure out where in the genome each clone belongs. Next they cut each clone into smaller, overlapping pieces the right size for sequencing—about 500 BP each. Finally, they sequence the pieces and use the overlaps to reconstruct the sequence of the whole clone.

The other strategy, called **"whole-genome shotgun"** method, involves breaking the genome up into small pieces, sequencing the pieces, and reassembling the pieces into the full genome sequence. Each of these approaches has advantages and disadvantages. The clone-by-clone method is reliable but slow, and the mapping step can be especially time-consuming. By contrast, the whole-genome shotgun method is potentially very fast, but it can be extremely difficult to put together so many tiny pieces of sequence all at once.

Both approaches have already been used to sequence whole genomes. The whole-genome shotgun method was used to sequence the genome of the bacterium *Haemophilus influenzae*, while the genome of baker's yeast, *Saccharomyces cerevisiae*, was sequenced with a clone-by-clone method. Sequencing the human genome was done using both approaches.

SEQUENCING METHODS

There are essentially two ways to sequence a genome. The BAC-to-BAC method, the first to be employed in human genome studies, is slow but sure. The BAC-to-BAC approach, also referred to as the map-based method, evolved from procedures developed by a number of researchers during the late 1980s and 90s and that continues to develop and change.

The other technique, known as whole genome shotgun sequencing, brings speed into the picture, enabling researchers to do the job in months to a year. The shotgun method was developed by GNN president J. Craig Venter in 1996.

Two approaches have been used to sequence the genome. They differ in the methods they use to cut up the DNA, assemble it in the correct order, and whether they map the chromosomes before decoding the sequence. First there was the BAC to BAC approach. A second, newer method is called whole genome shotgun sequencing.

Clone-by-Clone or BAC to BAC method

The BAC to BAC approach first creates a crude physical map of the whole genome before sequencing the DNA. Constructing a map requires cutting the chromosomes into large pieces and figuring out the order of these big chunks of DNA before taking a closer look and sequencing all the fragments.

Several copies of the genome are randomly cut into pieces that are about 150,000 base pairs (bp) long. Each of these 150,000 bp fragments is inserted into a BAC-a bacterial artificial chromosome. A BAC is a man made piece of DNA that can replicate inside a bacterial cell. The whole collection of BACs containing the entire human genome is called a BAC library, because each BAC is like a book in a library that can be accessed and copied. These pieces are fingerprinted to give each piece a unique identification tag that determines the order of the fragments. Fingerprinting involves cutting each BAC fragment with a single enzyme and finding common sequence landmarks in overlapping fragments that determine the location of each BAC along the chromosome. Then overlapping BACs with markers every 100,000 bp form a map of each chromosome. Each BAC is then broken randomly into 1,500 bp pieces and placed in another artificial piece of DNA called M13. This collection is known as an M13 library. All the M13 libraries are sequenced. 500 bp from one end of the fragment are sequenced generating millions of sequences. These sequences are fed into a computer program called PHRAP that looks for common sequences that join two fragments together.

Whole Genome Shotgun Sequencing

The shotgun sequencing method goes straight to the job of decoding, bypassing the need for a physical map. Therefore, it is much faster. Multiple copies of the genome are randomly shredded into pieces that are 2,000 base pairs (bp) long by squeezing the DNA through a pressurized syringe. This is done a second time to generate pieces that are 10,000 bp long. Each 2,000 and 10,000 bp fragment is inserted into a plasmid, which is a piece of DNA that can replicate in bacteria. The two collections of plasmids containing 2,000 and 10,000 bp chunks of human DNA are known as plasmid libraries. Both the 2,000 and the 10,000 bp plasmid libraries are sequenced. 500 bp from each end of each fragment are decoded generating millions of sequences. Sequencing both ends of each insert is critical for the assembling the entire chromosome. Computer algorithms assemble the millions of sequenced fragments into a continuous stretch resembling each chromosome.

Assembling the Genome

Genome assembly is the job of computer programs known, appropriately enough, as "assemblers." These programs work by finding and analyzing overlaps or identical DNA sequences at either end of two different reads. On first sight one might think that reads that overlap belong next to each other in the final genome sequence. However, the genome contains over 30 percent of sequence that is repeated several times, so that a repeat overlap might also occur between fragments that are millions of base pairs apart in the genome.

The task of the assembler is to compare each read to every other, then to put all the reads in the proper order based on how they overlap, not using the repeat overlaps. The outcome of an assembly is a collection of big stretches of the genome that are put together correctly. Assembler programs have continuously improved since the first such software was written in the early 1980s. More powerful computers are also helping scientists assemble larger pieces of DNA faster than ever before.

Nevertheless, even the most powerful assembly software relies more on elegance and simplicity than on brute force. Many assemblers are only a fraction of the size of a typical word processing program—150,000 to 200,000 lines of code as opposed to several million. A program used to assemble the human genome would fit easily on the hard disk of a typical personal computer. When a genome has never been sequenced before, there is nothing to tell its explorers whether they have sequenced it correctly. When a genome has never been sequenced before, there is nothing to tell its explorers whether they have sequenced it correctly.

Errors can emerge at any stage of the process—when DNA is chopped up, when it is copied, as it goes through the sequencing machine, or as it is put together. Some sequences are particularly difficult to copy or to sequence and get left out. And random "noise" in the data can cause a base to be misidentified or overlooked. But a combination of redundancy and careful checking helps ensure that errors in genome sequencing are kept to an absolute minimum.

One trick for eliminating errors is to sequence the genome more than once. That is, scientists chop multiple copies of a genome up in such a way that each base is sequenced several times—6 to 10 times on average, depending on the specific project. That way, if the sequencing machine gets a base wrong, or if a piece of DNA slips through the cracks and doesn't get sequenced, there are likely to be other, correct reads that will provide the sequence.

In addition to identifying DNA bases, software on automatic sequencing machines can evaluate the probability that a base really is the base it appears to be. Error probabilities for all of the bases in a read are added together for an estimate of the number of errors in the sequence.

Bad reads or parts of reads—those with a lot of errors or question marks—are weeded out before they even make it to the assembly stage. With slab-gel machines some of this quality control is done by humans, while with capillary sequencers it is exclusively the province of computers. In addition, assembler software compares all the different reads that cover the same stretch of DNA and generates what is known as a "consensus" sequence. For example, if a certain base comes out as an A nine times and C the tenth, then chances are the base is really an A. An assembler is designed to sift through conflicting information and decide which sequence is likely to be right.

Once a sequence is assembled, there are several ways to make sure it has been put together correctly. The sequence may be checked against small parts of the genome that have previously been sequenced and assembled or against various landmarks on genome maps. In other words, if an assembly is consistent with scattered bits of known information, that is a good sign it is correct overall. Although computer programs can help resolve gaps and uncertainties in a genome sequence, much of the final polishing is still done by people known as finishers. These expert workers identify gaps in the sequence, design experiments to fill in those gaps, and determine how to collect any additional information that is necessary.

There is no mechanical substitute for the intuition and intelligence of an experienced finisher, so finishing is currently a bottleneck in the process of DNA sequencing. Automatic sequencing machines can churn out raw sequence much faster than humans can analyze and polish these sequences. Many scientists foresee a day when genome sequencing will be routine—when sequencing the genomes of many different species will help biologists understand the patterns of evolution, or when sequencing the genomes of individual humans will help doctors design tailor-made medicine. But until speedy machines become finishers as well as sequencers, that scenario will remain science fiction.

HUMAN GENOME

The human genome is a lot bigger than other genomes that have been sequenced in the past. Most genomes that have been sequenced to date belong to viruses, bacteria, or other simple forms of life with relatively small genomes. The human genome is about a thousand times larger than an average bacterial genome. Even the fruit fly genome, the largest genome sequenced prior to the human genome, is just 165 million base pairs—less than a tenth the size of the human genome. In addition, the human genome is about 25 to 50 percent repetitive DNA, but bacterial and viral genomes contain very little of this exasperating stuff. In repetitive DNA, the same short sequence is repeated over and over again. For example, somewhere in the genome the sequence ATG may be repeated 150 times in a row; elsewhere there may be 40 consecutive copies of the sequence CCTTGCT.

In jigsaw puzzle terms, a genome with a lot of repetitive DNA would be like a puzzle that includes a large number of identical or near-identical pieces—one in which the entire foreground is a featureless field of small, pink flowers, for example.

Like repetitive jigsaw puzzles, repetitive DNA can be difficult to assemble. It is often difficult for scientists to determine how much repetitive sequence belongs where. For example, 100 copies of ATG may belong in one spot in the genome, or it may be that only 60 copies belong there and 40 copies belong somewhere else. Repetitive DNA may also be more difficult to sequence than other DNA. Sometimes the procedures used to copy DNA and prepare it for sequencing do not work on repetitive DNA, and a sequencing machine may have a hard time reading the same string of letters over and over.

Future of Human Genome sequence

The working-draft DNA sequence and the more polished 2003 version represent an enormous achievement, akin in scientific importance, some say, to developing the periodic table of elements. And, as in most major scientific advances, much work remains to realize the full potential of the accomplishment. Early explorations of the human genome, now joined by projects on the genomes of several other organisms, are generating data whose volume and complex analyses are unprecedented in biology. Genomic-scale technologies will be needed to study and compare entire genomes, sets of expressed RNAs or proteins, gene families from a large number of species, variation among individuals, and the classes of gene regulatory elements.

Deriving meaningful knowledge from DNA sequences will define biological research through the coming decades and require the expertise and creativity of teams of biologists, chemists, engineers, and computational scientists, among others. A sampling follows of some research challenges in genetics--what we still don't know, even with the full human DNA sequence in hand.

- Gene number, exact locations, and functions
- Gene regulation.
- DNA sequence organization.
- Chromosomal structure and organization.
- Noncoding DNA types, amount, distribution, information content, and functions.

- Coordination of gene expression, protein synthesis, and post-translational events.
- Interaction of proteins in complex molecular machines.
- Predicted vs experimentally determined gene function.
- Evolutionary conservation among organisms.
- Protein conservation (structure and function).
- Proteomes (total protein content and function) in organisms.
- Correlation of SNPs (single-base DNA variations among individuals) with health and disease.
- Disease-susceptibility prediction based on gene sequence variation.
- Genes involved in complex traits and multigene diseases.
- Complex systems biology, including microbial consortia useful for environmental restoration.
- Developmental genetics, genomics

RESTRICTION MAPPING OF THE DNA

Restriction mapping is the process of obtaining structural information on a piece of DNA by the use of restriction enzymes.

Restriction enzymes

Restriction enzymes are enzymes that cut DNA at specific recognition sequences called "sites." They probably evolved as a bacterial defense against DNA bacteriophage. DNA invading a bacterial cell defended by these enzymes will be digested into small, non-functional pieces. The name "restriction enzyme" comes from the enzyme's function of restricting access to the cell. A bacterium protects its own DNA from these restriction enzymes by having another enzyme present that modifies these sites by adding a methyl group. For example, *E.coli* makes the restriction enzyme Eco RI and the methylating enzyme Eco RI methylase. The methylase modifies Eco RI sites in the bacteria's own genome to prevent it from being digested.

Restriction enzymes are endonucleases that recognize specific 4 to 8 base regions of DNA. For example, one restriction enzyme, Eco RI, recognizes the following six base sequence:

5' . . . G-A-A-T-T-C . . . 3'

3' . . . C-T-T-A-A-G . . . 5'

A piece of DNA incubated with Eco RI in the proper buffer conditions will be cut wherever this sequence appears. As you can see, this site is palindromic; that is, reading the upper strand from 5' to 3' is the same as reading the lower strand from 5' to 3'. As a result, each strand of the DNA can self-anneal and the DNA forms a small cruciform structure:

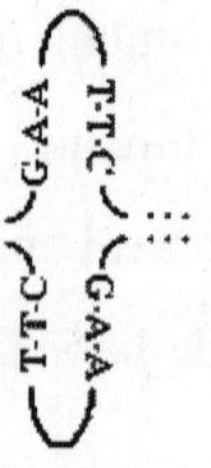

All restriction enzyme sites are palindromic. This structure may help the enzyme to recognize the sequence that it is designed to cut. There are hundreds of restriction enzymes that have been isolated and each one recognizes its own specific nucleotide sequence. Sites for each restriction enzyme are distributed randomly throughout a particular DNA stretch. Digestion of DNA by restriction enzymes is very reproducible; every time a specific piece of DNA is cut by a specific enzyme, the same pattern of digestion will occur. Restriction enzymes are commercially available and their use has made manipulating DNA very easy.

Restriction Mapping

Restriction mapping involves digesting DNA with a series of restriction enzymes and then separating the resultant DNA fragments by agarose gel electrophoresis. The distance between restriction enzyme sites can be determined by the patterns of fragments that are produced by the restriction enzyme digestion. In this way, information about the structure of an unknown piece of DNA can be obtained.

Uses of Restriction Mapping

Restriction map information is important for many techniques used to manipulate DNA. One application is to cut a large piece of DNA into smaller fragments to allow it to be sequenced. Genes and cDNAs can be thousands of kilobases long (megabases - Mb); however, they can only be sequenced 400 bases at a time. DNA must be chopped up into smaller pieces and subcloned to perform the sequencing. Also, restriction mapping is an easy way to compare DNA fragments without having any information of their nucleotide sequence (Fig. 6.3).

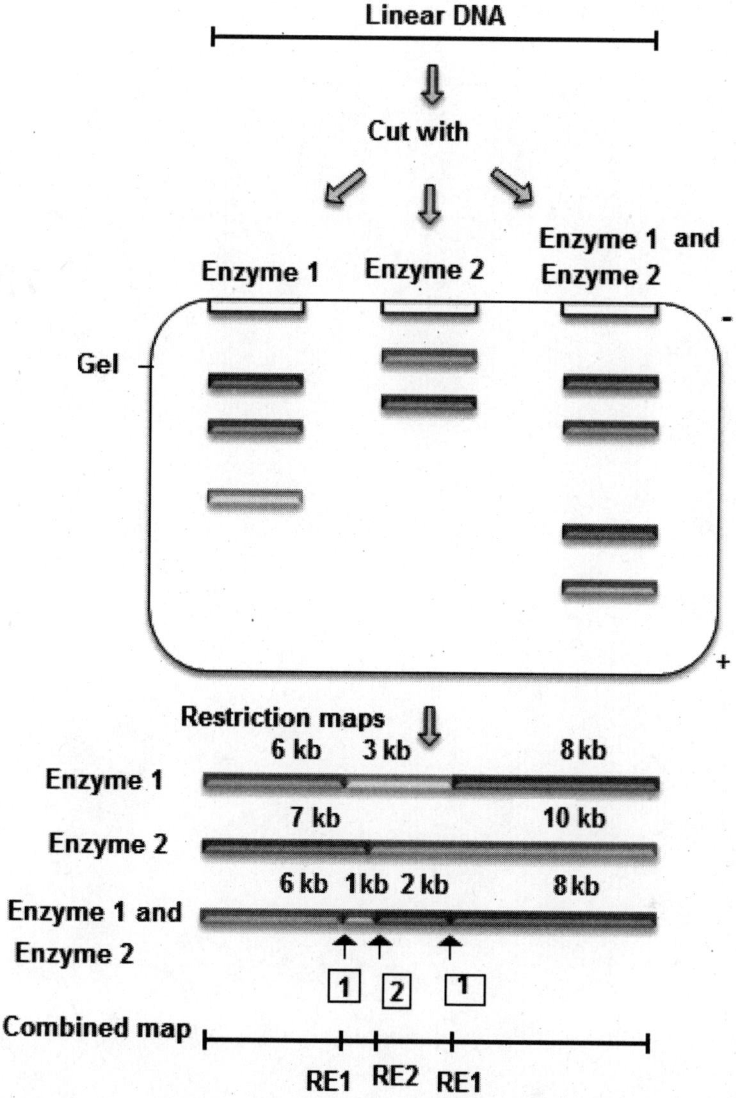

Fig. 6.3 Restriction mapping

7

cDNA Synthesis and Cloning

Complementary DNA (cDNA) is DNA synthesized from a mature mRNA template in a reaction catalyzed by the enzyme reverse transcriptase. cDNA is often used to clone eukaryotic genes in prokaryotes. When scientists want to express a specific protein in a cell that does not normally express that protein (i.e., heterologous expression), they will transfer the cDNA that codes for the protein to the recipient cell. cDNA is also produced by retroviruses (such as HIV-1, HIV-2, Simian Immunodeficiency Virus, etc.) which are integrated into its host to create a provirus.

The central dogma of molecular biology outlines that in synthesizing proteins, DNA is transcribed into mRNA, which is translated into protein. One difference between eukaryotic and prokaryotic genes is that eukaryotic genes can contain introns (intervening sequences), which are not coding sequences, and must be removed from the RNA primary transcript before it becomes mRNA and can be translated into protein. Prokaryotic genes have no introns, so their RNA is not subject to splicing.

Often it is desirable to express eukaryotic genes in prokaryotic cells. A simplified method of doing so would include the addition of eukaryotic DNA to a prokaryotic host, which would transcribe the DNA to mRNA and then translate it to protein. However, as eukaryotic DNA has introns, and since prokaryotes lack the machinery to splice them, the splicing of eukaryotic DNA must be done prior to adding the eukaryotic DNA into the host. This DNA which was made as a complementary to the RNA is called complementary DNA (cDNA). To obtain expression of the protein encoded by the eukaryotic cDNA, prokaryotic regulatory sequences would also be required (e.g. a promoter).

Complementary DNA is often used in gene cloning or as gene probes or in the creation of a cDNA library. When scientists transfer a gene from one cell into another cell in order to express the new genetic material as a protein in the recipient cell, the cDNA will be added to the recipient (rather than the entire gene), because the DNA for an entire gene may include DNA that does not code for the protein or that interrupts the coding sequence of the protein (e.g., introns). Partial sequences of cDNAs are often obtained as expressed sequence tags. Some viruses also use cDNA to turn their viral RNA into mRNA (viral RNA → cDNA → mRNA). The mRNA is used to make viral proteins to take over the host cell.

cDNA LIBRARY

A cDNA library is a collection of cloned cDNA (complementary DNA) fragments inserted into a collection of host cells, which together constitute some portion of the transcriptome of the organism. cDNA is produced from fully transcribed mRNA found in the nucleus and therefore contains only the expressed genes of an organism. Similarly, tissue specific cDNA libraries can be produced. In eukaryotic cells the mature mRNA is already spliced, hence the cDNA produced lacks introns and can be readily expressed in a bacterial cell. While information in cDNA libraries is a powerful and useful tool since gene products are easily identified, the libraries lack information about enhancers, introns, and other regulatory elements found in a genomic DNA library.

cDNA Library Construction

cDNA is created from a mature mRNA from a eukaryotic cell with the use of an enzyme known as reverse transcriptase. In eukaryotes, a poly-(A) tail (consisting of a long sequence of adenine nucleotides) distinguishes mRNA from tRNA and rRNA and can therefore be used as a primer site for reverse transcription.

mRNA enrichment

Most eukaryotic mRNAs are polyadenylated at the 3′end to form a poly(A) tail. This has an important practical consequence that has been exploited by molecular biologists. The poly(A) region can be used to selectively isolate mRNA from total RNA by affinity chromatography (Fig.. 7.1) . The purified mRNA can then be used as a template for synthesis of a complementary DNA. Total RNA is extracted from a specific cell type that expresses a specific set of genes. Of this total cellular RNA, 80–90% is rRNA, tRNA, and histone mRNA, not all of which have a poly(A) tail. These RNAs can be separated from the poly(A) mRNA by passing the total RNA through an affinity column of oligo(dT) or oligo(U) bound to resin beads. Under conditions of relatively high salt the poly(A) RNA is retained by formation of hydrogen bonds with the complementary bases, and the RNA lacking a poly(A) tail flows through. The salt conditions for hybridization are similar to the ion concentration in cells (e.g. 0.3–0.6 M NaCl). The poly(A) mRNA is then eluted from the column in low salt elution buffer (e.g. 0.01 M NaCl), which promotes denaturation of the hybrid.

First strand synthesis of cDNA

A number of strategies can be used to synthesize cDNA from purified mRNA. One strategy is as follows. In brief, cDNA is synthesized by the action of reverse transcriptase and DNA polymerase. The reverse transcriptase catalyzes the synthesis of a single-stranded DNA from the mRNA template. Like a regular DNA polymerase, reverse transcriptase also needs a primer to get started. A poly(dT) primer is added to provide a free 3′-OH end that can be used for extension by reverse transcriptase in the presence of deoxynucleoside triphosphates (dNTPs). Usually a viral reverse transcriptase is employed such as one from avian myeloblastosis virus (AMV). The reverse transcriptase adds dNTPs from 5′ to 3′ by complementary base pairing. This is called first strand synthesis. The mRNA is then degraded with a ribonuclease or an alkaline solution.

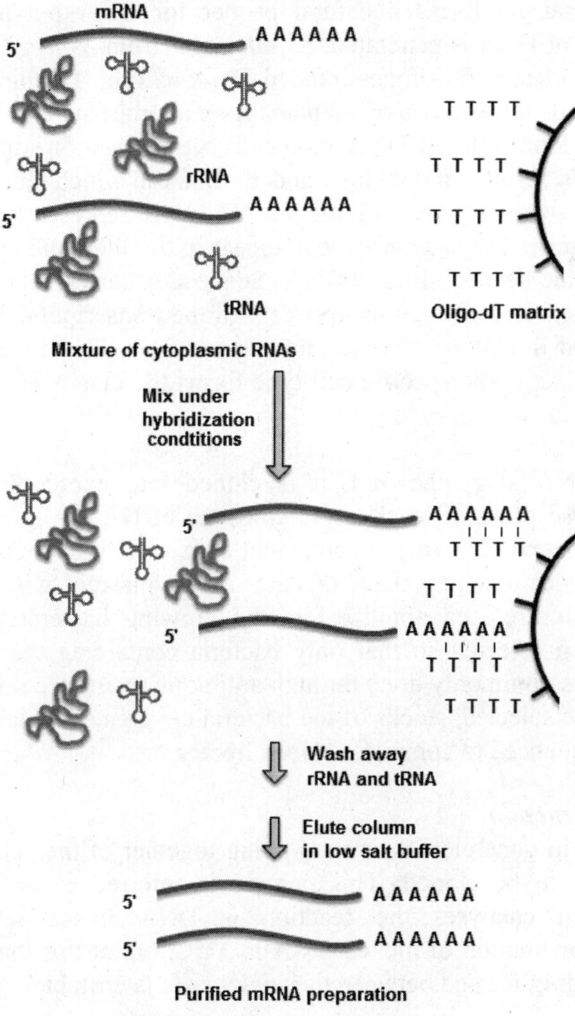

Fig. 7.1 mRNA enrichment

Second strand synthesis of cDNA

For most applications, including cloning of cDNA, double stranded DNA is required. The second DNA strand is generated by the Klenow fragment of DNA polymerase from *E.coli*. The 5'→ 3' exonuclease activity of DNA polymerase I from *E. coli* makes it unsuitable for many applications. However, this enzymatic activity can be readily removed from the holoenzyme by exposure to a protease. The large or Klenow fragment of DNA polymerase I generated by proteolysis has 5′ → 3′ polymerase and 3′ → 5′ exonuclease (proofreading) activity, and is widely used in molecular biology. Commercially available Klenow fragments are usually produced by expression in bacteria from a truncated form of the DNA polymerase I gene. There is a tendency for the reverse transcriptase enzyme used in first strand synthesis to loop back on itself and start to make another complementary strand. This hairpin forms a natural primer for DNA polymerase and a second strand of DNA is generated. S1 nuclease (from *Aspergillus oryzae*) is then added to cleave the single-stranded DNA hairpin. Double-strand DNA linkers with ends that are complementary to an appropriate cloning vector are added to the double-strand DNA molecule before ligation into the cloning vector. The end result is a double-stranded cDNA in which the second strand corresponds to the sequence of the mRNA, thus representing the coding strand of the gene. The sequences that appear in the literature are the 5′ → 3′ sequences of the second strand cDNA. Sequences corresponding to introns and to promoters and all regions upstream of the transcriptional start site are not represented in cDNAs. The library created from all the cDNAs derived from the mRNAs in the specific cell type forms the cDNA library of cDNA clones (Fig. 7.2).

After the cDNA is synthesized, it is cloned into expression vectors or plasmids. These plasmids each containing one cDNA are transformed into bacterial competent cells (made competent by a variety of chemical or other methods in order to allow them to take up the plasmid with the cDNA). These plasmids are then amplified in the growing bacteria. The bacteria clones are then selected so that only bacteria containing the plasmid will survive. This is commonly done through antibiotic resistance selection. Once the bacteria are selected, stocks of the bacteria are created which can later be grown and sequenced to compile a cDNA library.

Library Construction

The next step in gene cloning is the joining together of the vector molecule and the DNA to be cloned. This process is referred to as ligation. The enzyme which catalyses the reaction is DNA ligase. The optimum temperature for ligation of nicked DNA is 37°C, but at this temperature the hydrogen bonding formed between the sticky ends is unstable.

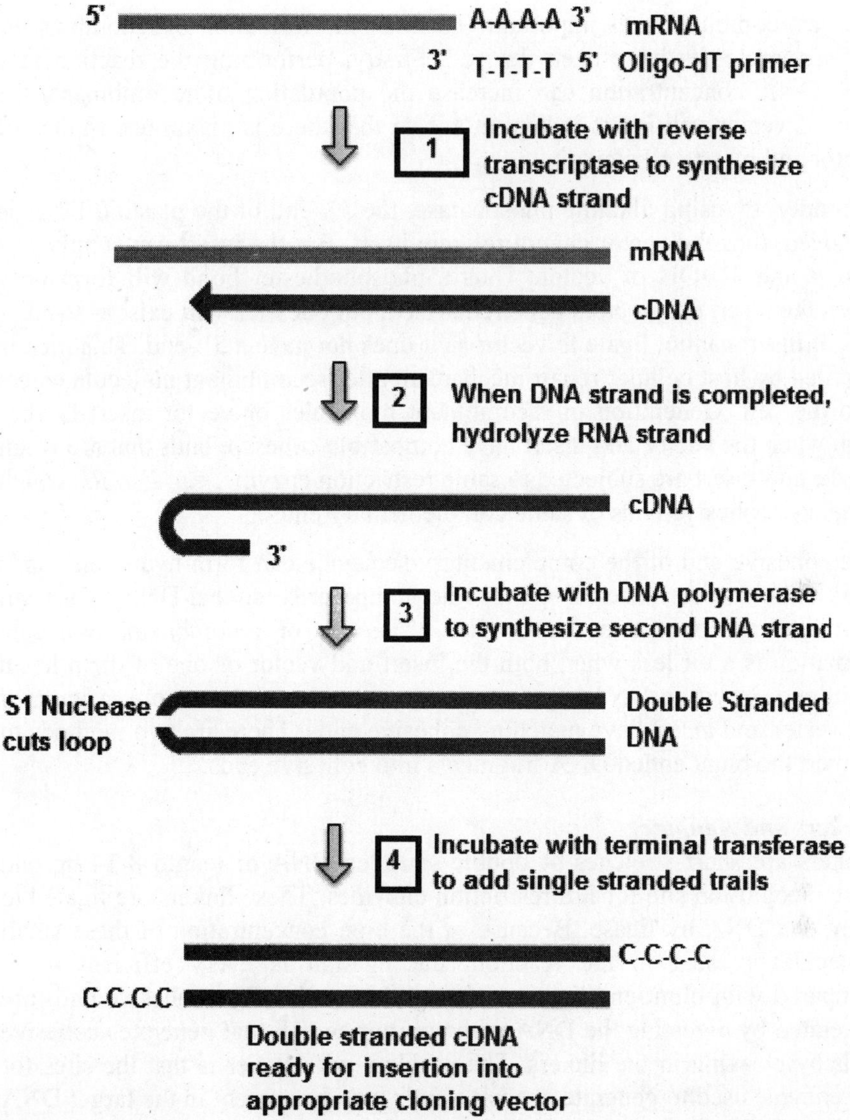

Fig. 7.2 cDNA synthesis

Hence, the ligation reaction is carried out at 4°C. Even though the rate of reaction is slow and takes a long time to complete, this temperature is preferred, as the success rate is high. The ligation reaction of vector and inserts generate 3 types of products

1) The insert gets circularized without ligating to vector.

2) The vector recircularizes without insert and

3) The vector-insert gets ligated.

The last combination is important. To favor the formation of recombinants, various strategies have been devised. Firstly, performing the reaction at a high DNA concentration can increase the population of recombinant, the ratio of vector and insert is kept at 1:2 so that there is maximum chance of vector and insert collision.

Secondly, by using alkaline phosphatase, the 5' -end of the plasmid DNA is removed, thus the vector cannot religate itself. But the insert can supply 5' -PO_4 to the 3' -OH of vector. Thus a phosphodiester bond will form only when the insert and vectors get circularized, but one nick will exist as the 3 ' -OH of insert cannot ligate to vector as it does not have a 5' -end. This nick is repaired by host cellular repair mechanism after recombinant molecule enters into the cell. Generation of recombinant molecules or vector insert is very high when the vector and insert have compatible cohesive ends that are when vector and insert are subjected to same restriction enzyme, e.g. *Eco RI,* which generates cohesive ends of same complementary ends.

The cohesive end of the complementary sequence can form hydrogen bonds and can hold vector insert DNA together temporarily, so that DNA ligase can form the phosphodiester bond. The efficiency of recombinant molecule formation is a bit less when both the insert and vector or one of them has a blunt end. Inserting DNA segments into vectors are always more efficient if the vector and insert have matching cohesive ends. There are two methods to convert the blunt ended DNA fragments into cohesive ends.

Linkers and Adapters

Linkers are short stretches of double stranded DNA of length 8-14 bp and have recognition site for 3-8 restriction enzymes. These linkers are ligated to blunt end DNA by ligase. Because of the high concentration of these small molecules present in the reaction, the ligation is every efficient when compared with blunt-end ligation of large molecules. The cohesive ends are generated by digesting the DNA with appropriate RE that generates cohesive ends by cleaving in the linkers. The problem with linker is that the sites for the enzyme used to generate cohesive ends may be present in the target DNA fragment. This drawback limits the use of linkers for cloning (Fig. 7.3).

Adapters are linkers with cohesive ends or a linker digested with RE, before ligation. The most widely used definition is cut linkers also called as adapters. They are not perfectly double stranded non single stranded. By adding adaptors to the ends of a DNA, sequences that are blunt can be converted into cohesive ends.

Homopolymer OR T/A Tailing

This method uses the ability of annealing of complementary strands or sequences. Suppose a vector has an oligo (dA) sequence at the 3'-OH end and the insert has an oligo (dT) sequence at its 3'-OH end. Then when both the molecules are mixed, the molecules are held by hydrogen bond or can anneal until the ligase joins them by phosphodiester bond. The important component in this method is terminal deoxynucleotidyl transferase. This enzyme adds nucleotides at the 3'-OH end of DNA without any complementary sequence. It can add up to 10-40 homopolymer residues at the end (Fig. 7.4). Commonly, the annealed circles with nicks and mismatching are used directly for transformation as these mistakes are repaired after the recombinant molecule enters into the host.

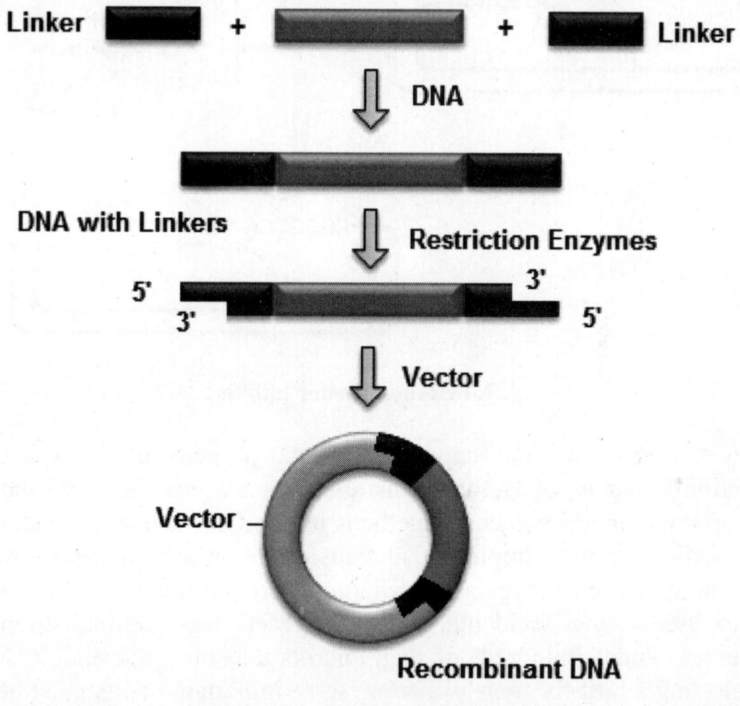

Fig. 7.3 Cloning using Linkers

Transformation and Multiplication of Clones

The central step in a gene cloning procedure is to transfer a recombinant clone generated in vitro, into bacteria or any other host. The concept and feasibility of molecular cloning is centered around two principles. Ligation in vitro generally yields a population of DNA molecules out of which only some are important. Hence the transformation step should ensure that one cell receives a single plasmid or molecule. This results in separation of each recombination from all the others. Each recipient cell needs to separate from all the others in the population to permit isolation of a clone of cells.

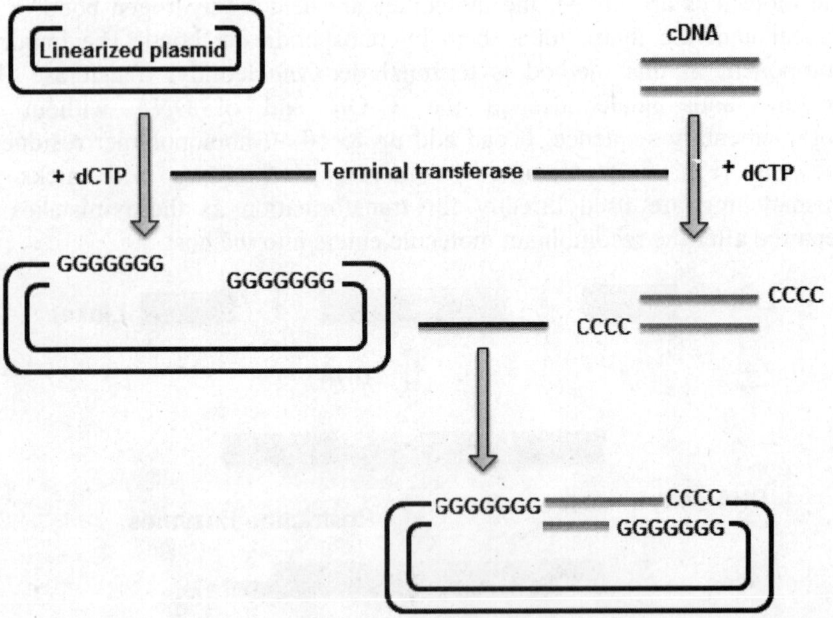

Fig. 7.4 Homopolymer tailing

Isolation of a clone of cells depends upon the property of the host-vector combination being used. Before transformation we have to keep the host vector combination and screening methods in mind and select the host, as the host properties are very important in transformation and multiplication. In most of the cases, microorganisms which are proved unlikely to survive in nature are used so that accidental release of genetically modified strains can be prevented. Today, hundreds of such microorganisms are available. Hence when selecting a host for transformation, three important points must be kept in mind. **Firstly** the host must supply the factor required for the vector replication and should not contain any elements that inhibit vector multiplication or prevent some screening methods being used. **Secondly**, the cell should not contain any active restriction enzymes being synthesized, as this will cleave the recombinant molecule. Even the host cells which provide *E. coli* dam and dam methyl transferase are avoided, as they will generate replicated recombinant molecules which cannot be cleaved by some restriction enzymes due to methylation of the recognition site.

Lastly, the host should not have any phenotypic character which is similar to the vector. For example, if the vector codes for ampicillin resistance, the host must be susceptible to ampicillin antibiotic in the absence of vector. As a general rule, the host cell should be sensitive to particular antibiotics or toxins and should not harbour any extraneous plasmid (e.g. F^+ *E. coli* cells). After the host is decided, the second step is deciding the approach to introduce the recombinant DNA molecule.

Transfection Methods

In organisms like bacteria and other microbes, or even in higher plants, the uptake of genes by cells is often described by the term 'transformation'. However in animals this term has been replaced by the term 'transfection', because the term 'transformation' in animal cell culture is used to describe phenotypic alteration of cells. The usage of the term 'transformation' for 'cell alteration' has been unfortunate and discontinuation of its usage for this purpose is suggested. Transfection or gene transfer in animals may be carried out at the cellular level to get transfected cells, which may be used for a variety of purposes such as production of chemicals and pharmaceutical drugs. It may also be undertaken for basic studies involving study of structure 'and function of genes. Although many mammalian cell lines have been regularly utilized for these purposes, transfection has also been achieved successfully for the production of transgenic animals.

Direct Gene Transformations

Development of gene-cloning vectors for plants aims at introducing foreign DNA into the plant using some natural system, viral or bacterial. When this is not possible, other methods are sought. Direct transformation implies that the cells take up the foreign gene of interest without the help of a vector. The gene vectors derived from Ti Plasmid have provided a means of testing the activity of recombinant selectable marker genes in plants. Once a particular recombinant gene has been shown to be active in plant cells and to confer a selectable advantage on the recipient cells, it can be used in experiments aimed at developing more direct ways of introducing isolated genes into plants.

Plant molecular biologists would like to apply the new methods of gene transfer to genetic modification of cereals, the major crops of agriculture. Research has been frustrated, however, because *Agrobacterium* apparently does not infect cereals and because attempts to regenerate whole plants from cereal protoplasts have generally failed. A number of approaches are currently being explored as alternative ways of transferring genes into cereal plants. These approaches include the uptake of DNA through the cell walls of germinating pollen, protoplasts, and the injection of DNA into germ cells before they undergo meiosis, liposome mediated transfer, and electroporation.

Direct Gene Uptake by Protoplasts

Protoplasts are cells without rigid cellulose walls. It has been shown that plant protoplasts treated with polyethylene glycol, commonly used to induce protoplast fusion, will take up DNA from their surrounding medium. More importantly, this DNA can then be stably integrated into the plant chromosomal DNA.

Direct DNA uptake by protoplasts can be stimulated by chemicals like polyethylene glycol (PEG). The technique is so efficient that virtually every protoplast system has proved transformable. PEG is also used to stimulate the uptake of liposomes and to improve the efficiency of electroporation. PEG at high concentration (15-25%) will precipitate ionic macromolecules such as DNA and stimulate their uptake by endocytosis without any gross damage to protoplasts. This is followed by cell wall formation and initiation of cell division. These cells can now be plated at low density on selection medium.

Initial studies using the above method were restricted to ***Petunia*** and ***Necotiana***. However, other plant systems (rice, maize, etc.) were also successfully used later. In these methods, PEG was used in combination with pure Ti plasmid, or calcium phosphate precipitated Ti plasmid mixed with a carrier DNA. Transformation frequencies upto 1 in 100 have been achieved by this method. Nevertheless, there are serious problems in using this method for getting transgenic plants, mainly due to difficulties encountered in plant regeneration from protoplasts.

Co cultivation Techniques

A further technique which would be useful when transferring DNA present in bacteria to higher plant cells involves the direct fusion of bacterial and plant cells. Its key components are:

1) Rapid culture method,
2) optimal conditions for co cultivation of *A. tumefaciens* cells with protoplasts, and
3) selection for hormone autotrophy.

Using the cocultivation technique adapted to the rapid plating system and early selection for hormone autotrophy, transformation frequencies as high as 80% have been observed with *A. tumefaciens* strains carrying octopine, nopaline or agropine type Ti plasmids.

Southern hybridization analysis showed the presence of T-DNA in several of the transformants. An important advantage of this system is that transformants can be identified by simply screening colonies for opine production in the absence of selective conditions.

Microprojectile Gun Method

To overcome the limitations of protoplast regeneration, high velocity microprojectiles are being used to deliver nucleic acids directly into intact plant cells or tissues.

During early 1990s, it was shown that DNA delivery to plant cells is also possible, when heavy **microparticles** (tungsten or gold) coated with the DNA of interest are accelerated to a very high initial velocity (1,400 ft per sec). These microprojectiles, each normally 1-3μm in diameter, are carried by a 'macroprojectile' or the '**bullet**' and are accelerated into living plant cells (target cells can be pollen, cultured cells, cells in differentiated tissues and meristems) so that they can penetrate cell walls of intact tissue. The acceleration is achieved either by an explosive charge (**cordite explosion**) or by using shock waves initiated by a high-voltage electric discharge (Fig. 7.5).

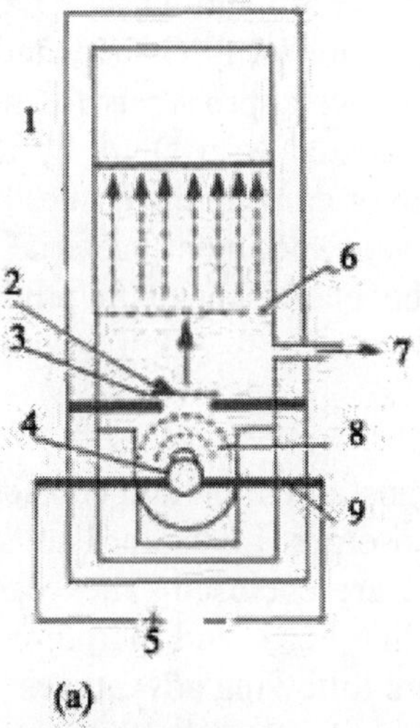

Fig. 7.5(a) Electrostatic Device

1. Target
2. Microprojectiles
3. Mylar Carrier Sheet
4. Water Droplet
5. High Voltage Discharge Device
6. Arrester Grid
7. Vacuum Support Plate
8. Sonic Shock Wave
9. Electrode

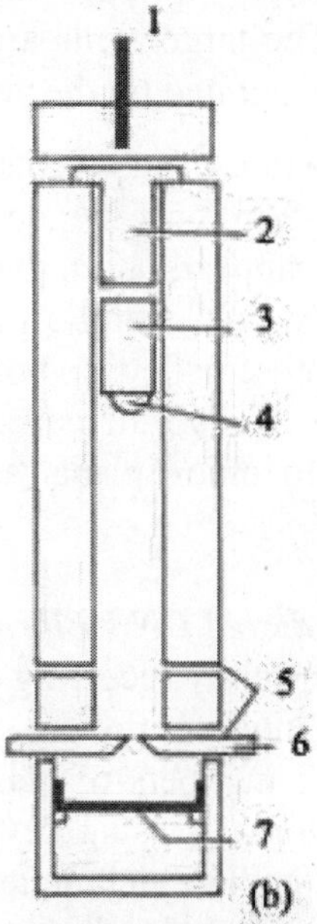

Fig. 7.5 (b) Ballistic Device

1. Firing Pin
2. Blank Charge
3. Nylon Microprojectile
4. Microprojectiles
5. Vents
6. Plate to Stop Nylon Projectile
7. Plate Containing target cells or tissue

In this method DNA is coated on the surface of tungsten particles which are projected by means of a particle gun into intact cells or tissues. The particles can penetrate through several layers of cells and can transform cells within tissue/explants. Soybean, tobacco, and maize have been transformed by this method. The particle gun is manufactured by DuPont. The DNA to be transformed is first coated on microbeads composed of tungsten or gold. The size of the beads ranges from 0.5 to 5.0 μm depending on the size and permeability of the target cells. A few microlitres of a slurry of these microprojectiles (DNA-coated tungsten particles or microbeads) are placed on the projectile carrier, which is accelerated to high velocity down the gun barrel in a particle gun apparatus. The microprojectile carrier is stopped by a plate at the end of the barrel while a small hole in the plate allows the microprojectiles to enter the target cells. These very dense particles thus acquire sufficient kinetic energy to penetrate intact cell walls and membranes. The target cells are placed in a partially evacuated chamber to expel the air generated by the velocity of the microprojectiles.

For transformation of monocots such as corn and wheat a particle gun could be useful, but for dicots the electroporation of PEG-mediated transformation of protoplasts appears attractive. Another approach is to shoot plant tissue with extremely fine metallic pellets coated with DNA by means of a 0.22 calibre gun. These pellets do not harm or damage the tissues but some deliver the DNA directly into the cells. In one experiment bacterial and viral genes were fired into onion plants and the plants began to produce the foreign proteins.

Liposome Mediated DNA Library

The system currently receiving the most attention and considered to hold the most promise involves the use of liposomes. Liposomes are small lipid bags, in which large numbers of plasmids are enclosed. They can be induced to fuse with protoplasts using devices like PEG, and therefore have been used for gene transfer. The technique offers following advantages: (i) protection of DNA/RNA from nuclease digestion, (ii) low cell toxicity, (iii) stability of nucleic acids due to encapsulation in liposomes, (iv) high degree of reproducibility and (v) applicability to a wide range of cell types.

In the experiments with tobacco protoplasts, TMV RNA was encapsulated in a variety of different liposome preparations using the reverse evaporation method. Incubation of tobacco protoplasts was done with negatively charged, phosphatidyl serine liposomes containing TMV RNA. Virus production could also be detected following incubations with neutral or positively changed liposome preparations. The same thing has been done with *Petunia* also. It is likely that liposomes will have their greatest application in transformation experiments utilizing non- Ti plasmid vectors and in experiments with protoplasts which are outside the host range of *A. tumefaciens*.

Uniform artificial lipid vesicles of 0.4 μm or more in diameter can be made by sonicating a mixture of phosphatidyl serine and an aqueous buffer. DNA or chromosomes were encapsulated in these large unilamellar vesicles. It was found that the infectivity of the SV40 DNA trapped in the liposomes and delivered to the cells was enhanced at least 100-fold over free D"I A added to the cells under the same conditions. Polyethylene glycol or glycerol treatment of the cells incubated with liposomes further increased their infectivity between 10- and 200-fold.

Microinjection and Macroinjection

Plant regeneration from transformed protoplasts, still remains a problem. Therefore cultured tissues, that encourage continued development of immature structures, provide alternate cellular targets for transformation. These immature structures may include immature embryos, meristems, immature pollen, isolated ovules, embryogenic suspension, cultured cells, etc. The main disadvantage of this technique is the production of chimeric plants with only a part of the plant transformed. However, from this plant transformed plants of single cell origin can be subsequently obtained. Utilizing this approach, transgenic chimeras were actually obtained in several crops including oilseed rape (*Brassica napus*).

When cells or protoplasts are used as targets in the technique of microinjection, glass micropipettes each with 0.5-10μm diameter tip are used for transfer of macromolecules into the cytoplasm or the nucleus of a recipient cell or protoplast (Fig. 7.6). The recipient cells are immobilized on a solid support (cover slip or slide, etc.) or artificially bound to a substrate or held by a pipette under suction (as done in animal systems).

Electroporation

Electroporation is a gene transfer method in which DNA is transferred into cells by using high voltage current for a fraction of second. This method can be applied to transfer DNA into plant cells, animal cells, yeast and bacteria. In this method, cells are placed between the electrodes present in an Electroporation chamber along with an ionic solution containing the vector DNA. An electric pulse generated by a capacitor is applied for about 10-45 milliseconds, which induces pores in the plasma membranes (Fig. 7.7). These pores appear to be round and are present for several minutes after the pulse during which vector DNA passes through the pores into the cell.

The transformation efficiency by this method depends upon the field strength or amount of current applied and the time of application. High voltage (1.5 kV) can be applied for a short duration of 10 milliseconds or a low voltage of 350 V can be applied for a longer duration of 54 milli seconds.

Transformation efficiency can also be increased by using various other methods such as adding 13% PEG solution into the electroporation chamber, linearing the DNA or heat shocking (45°C for 5 min.) prior to impulse.

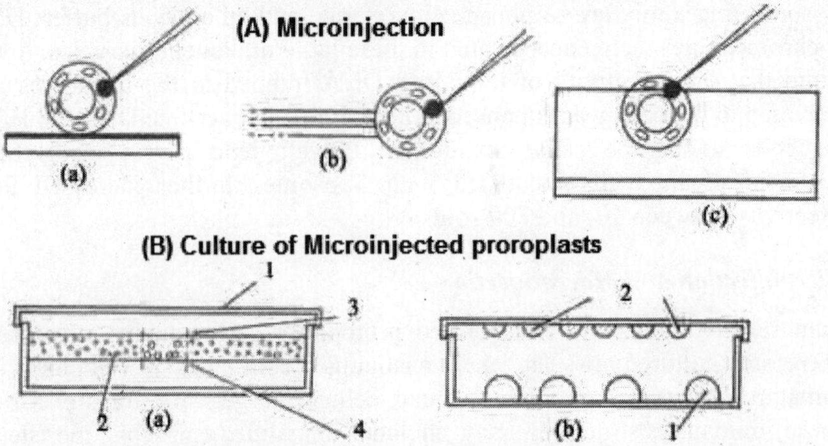

Fig. 7.6 (A) Microinjection: (a) Poly-L-Lysine (b) Holding pipette method (c) Agarose method

Fig. 7.6 (B) Culture of Microinjected Protoplasts: (a) Nylon Gauze chamber: 1 lid 2. Nurse culture Protoplasts suspended in liquid culture medium 3. Nylon gauze ring 4. Injected protoplasts (b) Hanging Droplets : 1.Larger droplet to prevent Desiccation 2.Droplets containing microinjected protoplasts

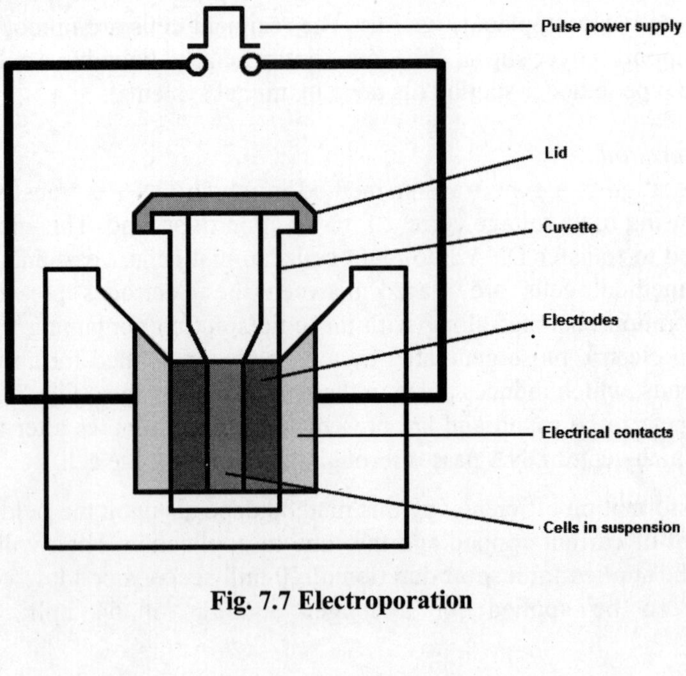

Fig. 7.7 Electroporation

Transformation Using Pollen or Pollen Tube

There has been a hope that DNA can be taken up by the germinating pollen and can either integrate into sperm nuclei or reach the zygote through the pollen tube pathway. Both these approaches have been tried and interesting phenotypic alterations suggesting gene transfer have been obtained. In no case, however, unequivocal proof of gene transfer has been available. In a number of experiments, when marker genes were used for transfer, only negative results were obtained. Several problems exist in this method and these include the presence of cell wall, nucleases, heterochromatic state of acceptor DNA, callose plugs in pollen tube, etc. Transgenic plants have never been recovered using this approach and this method, though very attractive, seems to have little potential for gene transfer.

Transformation by Ultrasonication

In wheat, tobacco and sugarbeet, explants after being cultured for a few days, were sonicated with plasmid DNA (carrying marker genes like *cat*, *nptII* and *gus*).When sonicated calli were transferred to selective medium, shoots were obtained, although all control calli (not sonicated with plasmid DNA) died. In tobacco, transgenic plants were obtained at a frequency of 22 per cent. This work was conducted at the Biotechnology Research Center at Beijing (China).

Increased Competence of E. coli by $CaCl_2$ Treatment

E. coli cells are generally poorly accessible to DNA molecules. But treatment with $CaCl_2$ makes them permeable to DNA; the process involved is poorly understood, Growing E.coli cells are isolated and suspended in 50 mM $CaCl_2$ at a concentration of 108-1010 cells/ml; the cells may be incubated for 12- 24 hr to increase the frequency of transformation. The recombinant DNA is then added; efficient transformation takes only a few minutes and the cells are plated on a suitable medium for the selection of transformed clones. The frequency of transformed cells is 106-107/µg of plasmid DNA; this is about one transformation per 10,000 plasmid molecules. This frequency can be further improved by using special *E. coli* strains, e.g., SK1590, SK1592, X1766, etc., and some specific conditions during transformation; these may raise the frequency to 5 X 108 transformed cells/µg of plasmid DNA. The transformed cells are suitably diluted and spread thinly on a suitable medium so that each cell is well separated and produces a separate colony. Generally, the medium is so designed also permit only the transformed cells to divide and produce colonies.

Infection by Recombinant DNAs Packaged as Virions

Alternatively, those recombinant DNA that have the λ phage cos sequences, e.g., those derived from cosmids, phasmids and λ vectors, are generally packaged in vitro into specially produced empty λ phage heads and complete λ particles are constituted. These phage particles are used to infect E. coli cells; this process is often called transfection. These recombinant DNAs can also be used to transform *E. coli* cells directly as naked DNA, using the $CaCl_2$ technique. Generally, transfection is far more efficient than direct transformation.

For example, the frequency of transfection by recombinant λ phage DNAs packaged in phage particles is up to 108 plaques/μg of DNA, while it is less than 103 plaques/μg DNA when the recombinant DNA is used for transformation by the $CaCl_2$ technique.

The infected/transformed bacterial cells are spread on a lawn of susceptible cells, where clear areas or plaques develop in the lawn. Plaques containing the recombinant DNA (A vector and phasmids) are identified and the phage particles collected from such plaques provide the purified vector/ recombinant DNA.

Screening of clones

Once we obtain a population of recombinant clones the next step is to identify a clone, which has the DNA insert of interest. The technique used for identification has to be highly precise and extremely sensitive to allow an accurate detection of a single clone from among the thousands obtained from a cloning experiment. The various strategies used for the purpose are briefly outlined below.

Colony Hybridization

This technique is used to identify those bacterial colonies in a plate, which contain a specific DNA sequence. These bacterial colonies are obtained from bacterial cells into which this sequence was introduced through genetic engineering, and the given sequence is represented by the probe used in the hybridization experiment (Fig. 7.8). The procedure for colony hybridization is briefly described below:

1) The bacterial cells subjected to transformation are plated onto a suitable agar plate; this is the master plate.
2) The colonies of master plate are replica plated onto a nitrocellulose filter membrane placed on agar medium. For replica plating, a block of wood or cork, of suitable diameter for the master plate, is covered with velvet cloth. This block is sterilized and then lowered into the master plate till the velvet touches all the colonies; the block is withdrawn and gently lowered onto the nitrocellulose filter so that the bacterial cells sticking on to the velvet are transferred onto the filter. The master plate is retained intact for later use. A reference point is marked both on the master plate and on the replica plate to facilitate later comparisons.

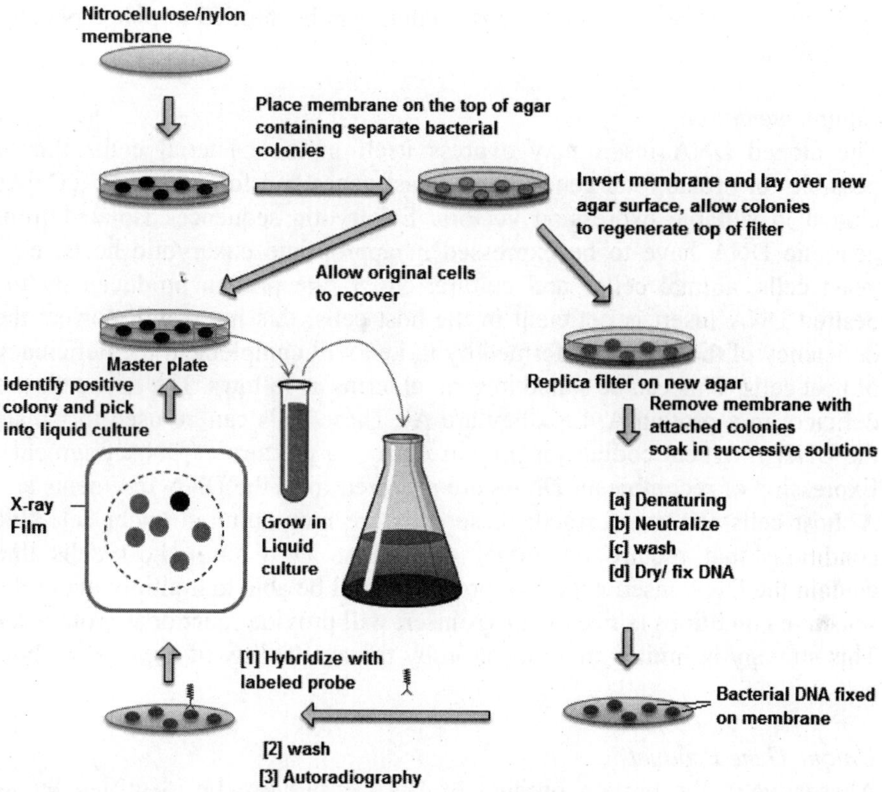

Fig. 7.8 Colony Hybridization

3) After the colonies appear, the filter is removed from the agar plate and treated with alkali to lyse the bacterial cells. This also denatures the DNA released from these cells.

4) The filter is treated with proteinase K to digest and remove the proteins; the denatured DNA remains bound to the filter.

5) The filter is now baked at 80° C to fix the DNA; this yields the DNA-print of bacterial colonies in the same relative positions as those of the colonies themselves in the master plate.

6) The filter is now hybridized with the radioactive probe; the probe represents the sequence of DNA segment used for transformation.. The unhybridized probe is removed by repeated washing.

7) The colonies whose DNA hybridizes with the probe are detected by autoradiography; only these colonies show up in the autoradiograph.

The positions of colonies showing up in the autoradiograph are compared with the master plate to identify the colonies that contain the DNA segment in question. The colonies are then picked up for further studies. A variation of this procedure can be applied to phage plaques as well.

Complementation

The cloned DNA insert may express itself in the bacterial cells; this is possible for prokaryotic genes, some yeast genes and for eukaryotic cDNAs cloned in suitable expression vectors. Eukaryotic sequences isolated from genomic DNA have to be expressed in appropriate eukaryotic hosts, e.g., yeast cells, animal cells and culture, etc. If the protein produced by the desired DNA insert is deficient in the host cells, this insert will correct the deficiency of the cells transformed by it, i.e., will complement the deficiency of host cells. This can be stated in general terms as follows. The host cells are deficient in a protein A, i.e., they are A^-. These cells can be used to isolate the DNA fragment coding for protein A from a mixture of DNA fragments. Expression of recombinant DNAs are prepared from the DNA fragments and $\mathbf{A}^-$ host cells are transformed; these cells are now cultured under selective conditions that require functional **A** product. Only those host cells that contain the DNA insert encoding protein **A** will be able to multiply under the selective conditions (since the DNA insert will provide functional protein **A**). This strategy is limited in application by the availability of appropriate host cells.

Unique Gene Products

Alternatively, the protein product of DNA insert can be identified by its unique function, i.e., a function not performed by the proteins of non transformed host cells. Such functions may relate to enzyme activities or hormone effects for which appropriate assays exist.

Antibodies Specific to the Protein Product

Finally, if the protein lacks a recognizable and measurable function, it can be detected by using specific antibodies. A practical approach is to divide the large number of recombinant clones into a convenient number of groups and to assay for the presence of the protein. The positive group is again divided into subgroups and assayed. In this manner, the positive groups are subdivided again and again till a single positive clone is identified. This approach is applicable to the previous strategy as well. The identification of proteins using antibodies may be achieved by western blotting, precipitation and electrophoresis or ELISA.

Colony/Plaque Screening with Antibodies

An efficient and rapid screening using antibodies is as follows. The antibody specific to the concerned gene product (i.e., protein) is spread uniformly over a solid support, e.g., plastic or paper disc, which is placed in contact with an agar layer containing lysed bacterial colonies or phage plaques. If any clone is producing the protein in question, it will bind to the antibody molecules present on the disc. The disc is removed from the agar, is treated with a second radiolabelled (generally with L25I) antibody, which is also specific to the same protein but in a region different from that recognized by the first antibody. These antibodies, therefore, will also bind to the protein molecule held by the first antibody; the location of radioactivity on the disc is determined by autoradiography. The colonies/plaques producing the protein are then identified and isolated from the master plate. This technique is analogous to colony hybridization and is able to screen large numbers of clones rather rapidly. But for this technique we require two different antibodies, which bind to two distinct domains of the desired protein, and this protein must not be produced by the non transformed host cells.

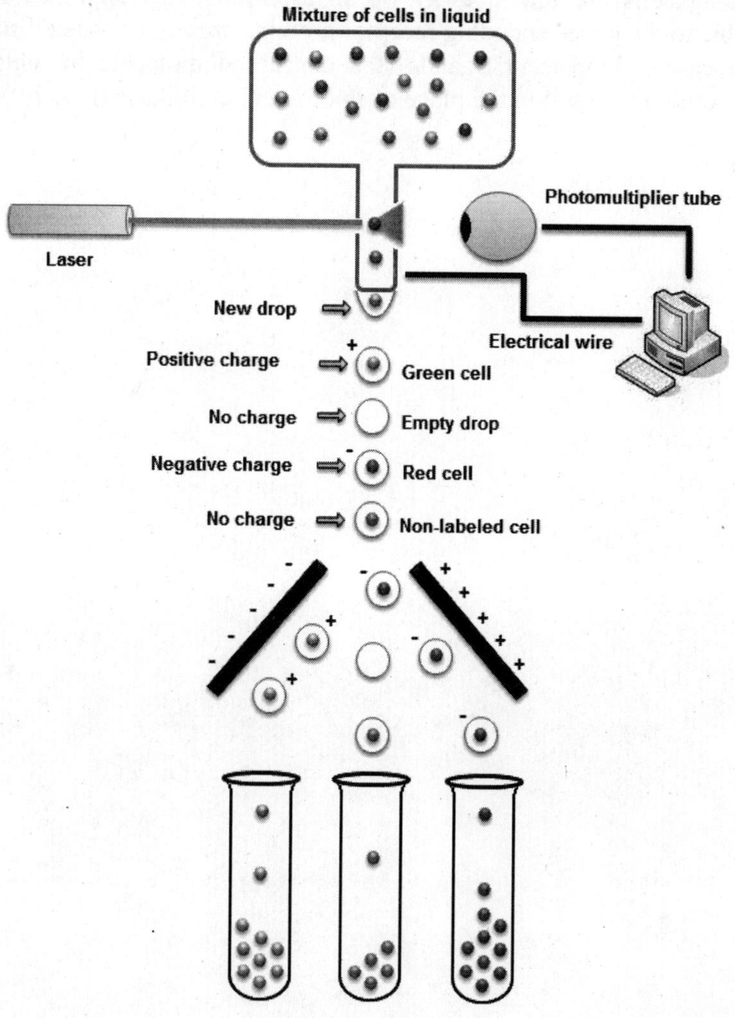

Fig. 7. 9 FACS

FACS

In case of animal cells, an automated system, called fluorescence activated cell sorter (FACS), can be used for very rapid (upto 1,000 cells/sec) sorting of transformed cells. This is applicable to all the genes whose products become arranged on the cell surface and are available for binding of specific antibodies. Therefore, these proteins must not be produced by the non transformed host cells. The antibody molecules are attached to a fluorescent molecule and the transformed cells are treated with this antibody specific for the desired protein. The cells containing on their surface the protein in question will interact with the fluorescent antibodies. Cells are then passed one by one in a stream between a laser and a fluorescence detector. The cells which fluoresce are deflected into a micro culture tray, while the non fluorescing cells are drawn away by an aspirator. This approach is also applicable to the genes encoding receptor proteins present on the cell surface; in such cases, fluorescent ligands (the concerned molecule to which the receptor binds) are used in the place of fluorescent antibodies (Fig. 7.9).

8

Alternative Strategies of Gene Cloning

GENE CLONING

If a gene from humans is placed in bacteria, it does not produce any protein or multiply, as bacteria does not recognize the gene as a gene. For the recognition and multiplication of the gene it has to carry some identification sequences or replicons. Such replicons are known as vectors or cloning vehicles. A composite DNA molecule formed by joining a gene or insert with a cloning vehicle is called as recombinant DNA or chimeric DNA or chimeras. The process of creating a chimeric DNA is called as gene cloning/ genetic engineering/gene manipulation or molecular cloning.

1) The first step in gene cloning is isolation of DNA from the organism being studied. Sometimes the DNA is made by using mRNA as reference or by chemical synthesis (PCR). Whatever be the method, the first step is to obtain the DNA molecule of interest.

2) The next step is to construct the chimeric DNA or recombinant DNA molecules by joining the vector or cloning vehicle with insert. Depending upon the host cell into which we transfer the insert, the vector or cloning vehicle varies. We use plasmids or phages for cloning in bacteria and SV 40 virus for cloning in animal cells.

3) The next step is transfer of recombinant molecules into the host so that it multiplies in the host. When the host cell divides, copies of the recombinant DNA molecules are passed on to the progeny.

4) The last step is identification of the bacteria which carries a recombinant molecule, from an array of bacteria with no recombinant molecules or having only a cloning vector.

5) After these initial steps, the protocols or procedures will diverge depending upon the goal of researchers. If they want to study the sequence of DNA, they proceed with DNA sequencing. If they want to make the protein of the gene they proceed for gene expression.

CLONING STRATEGIES

The complexity of the cloning experiment depends on the overall aim of the work and the type of source material from which the nucleic acids will be isolated for cloning. Thus, a strategy to isolate and sequence a relatively small DNA fragment from *E. coli* will be different from a strategy to produce a recombinant protein in transgenic eukaryotic organisms. Cloning strategies can be divided into two categories-those which involve the construction of gene bank or gene library and those require cloning of specific DNA fragments. For larger genomes of higher eukaryotes, it is necessary to clone the maximum number of different restriction fragments to obtain the gene of interest at a reasonable frequency. Thus, we obtain DNA sequences from large numbers of recombinants which contain a complete collection of nearly all DNA sequences in the entire genome. Such a collection of randomly cloned fragments that encompass the entire genome of a given species is called the gene library. Sometimes it is also called gene bank.

DIFFERENCE CLONING

It is a functional cloning approach that exploits differences in representation between DNA sources to isolate differentially expressed genes. It is much easier to isolate genes which are expressed in specific tissues. For instance, genes for storage proteins are expressed only in developing seeds, ova albumin gene is expressed in oviduct or globin gene is expressed in erythrocytes. Such genes can be easily isolated because mRNA extracted from these specific tissues will either exclusively belong to the gene of interest or it will be rich in this species of mRNA.

Other mRNA molecules in minor quantities can be eliminated, since these can be identified through hybridization with mRNA from tissues where this gene is silent. This strategy which is outlined in was actually followed for isolation and cloning of carrot genes expressed during the development of somatic embryos, and also for the isolation of several genes for storage proteins in crop plants. A number of genes with unknown gene products are expressed in specialized tissues. Mutations in these gene may also not result in easily detectable phenotypic changes. Isolation of such genes can be achieved through 'differential screening'. In this technique mRNA is prepared from contrasting tissues of plants such as the following:

(i) Tissues/plants exposed to different environmental conditions (e.g., drought).

(ii) Tissues, which differ in function.

(iii) Plants at different developmental stages.

In all these cases, two samples of mRNAs derived from two tissues or two plants should differ and the mRNA species which is not common in both the samples will be identified through differential screening using the following steps:

(i) mRNA is used for synthesis of cDNA using reverse transcriptase.

(ii) cDNA thus obtained from both samples of mRNA are labelled and used for sequential probing of a genomic DNA library.

(iii) Clones that are more strongly hybridized with the cDNA from one mRNA sample than with that from another contain genes that are differentially expressed in the samples (Fig.. 8.1).

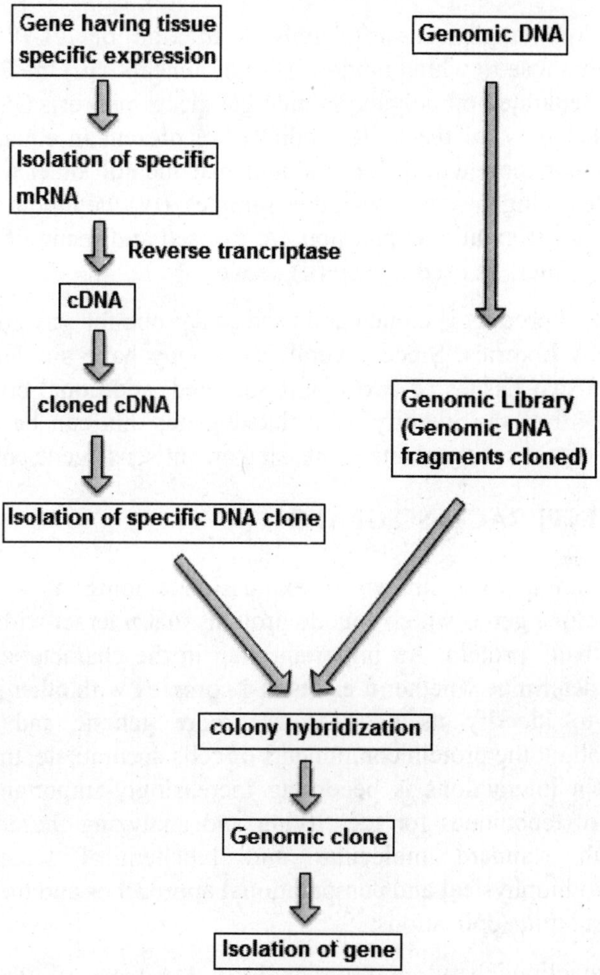

Fig. 8.1: Differential Cloning

One of the limitations of differential screening is that the techniques can be used for only those genes that have high expression, and may not prove suitable for those differentially expressed genes that have low expression (low abundance of mRNA). For such genes, another technique called 'differential display' is used.

The technique of differential display makes use of polymerase chain reaction (PCR) to amplify rare cDNAs. Following steps are involved isolate polyA mRNA form two sources as above (call them A and B) and run the following reactions separately with each of the two samples; (i) conduct RT-PCR, thus providing a means of differential amplification (ii) the RT-PCR product is amplified using a pair of primers, one of them being a random 10-mer and the other being the oligonucleotide used earlier in RT-PCR.

This will allow amplification of only a fraction of RT-PCR products, including only those to which random 10-mer anneals. (iii) the PCR products obtained are separated on polyacrylamide gel electrophoresis (PAGE); it will be noticed that most of the PCR products are present in equal amounts in both samples but some will differ in abundance and still other will be unique in one sample (being absent in the other sample); (iv) the unique bands or the bands showing differential expression are excised and reamplified using the same pair of primers as used in step (ii) above.

The reamplified product is cloned and used as a probe for screening cDNA or genomic DNA libraries. Since several genes may have similar patterns of expression in two tissues or two plants sampled, additional criteria will be needed to confirm the identify of isolated gene; this can be achieved by construction of transgenic plants expressing an antisense gene construct.

CLONING INTERACTING GENES

Interaction cloning (also known as expression cloning) is a technique to identify and clone genes which encode proteins that interact with a protein of interest, or "bait" protein. An important step in the characterization of any protein is to determine whether it exists in a complex with other proteins and, if it does, to identify its partners. As more genetic and biochemical information about the protein components of cells accumulate, the analysis of protein-protein interactions is becoming increasingly important. There are wide range of techniques for identifying and analyzing these interactions, starting with standard molecular and biochemical techniques, and progressing to biophysical and computational approaches and therapeutic and other post-genomic applications.

Specific interactions between proteins form the basis of many essential biological processes. Consequently, considerable effort has been made to identify those proteins that bind to proteins of interest. Typically, these interactions have been detected by using co-immuno-precipitation experiments in which antibody to a known protein is used to precipitate associated proteins as well. Such biochemical methods, however, result only in the identification of the apparent molecular mass of the associated proteins; obtaining cloned genes for these proteins is often a difficult process. These protein-protein interactions are critical to all cellular processes, and understanding them is key to understanding any biological system. One technique that can be used to study protein-protein interactions is the "yeast two hybrid" system.

Yeast Two Hybrid system

A protein is composed of modules or domains, which are individually folded units within the same polypeptide (protein) chain. The presence of these individual domains allow the same protein to perform different functions. The yeast two-hybrid technique uses two protein domains that have specific functions: a DNA-binding domain (BD), that is capable of binding to DNA (Fig. 8.2), and an activation domain (AD), that is capable of activating transcription of the DNA. Both of these domains are required for transcription, whereby DNA is copied in the form of mRNA, which is later translated into protein. In order for DNA to be transcribed, it requires a protein called a transcriptional activator (TA). This protein binds to the "promoter", a region situated upstream from the gene (coding region of the DNA) that serves as a docking site for the transcriptional protein. Once the TA has bound to the promoter, it is then able to activate transcription via its activation domain. Hence, the activity of a TA requires both a DNA binding domain and an activation domain. If either of these domains is absent, then transcription of the gene will fail.

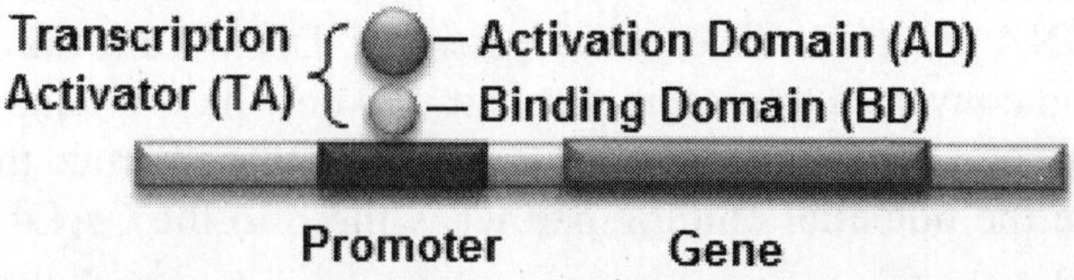

Fig. 8.2: Two domains of the Transcription factor

Furthermore, the binding domain and the activation domain do not necessarily have to be on the same protein. In fact, a protein with a DNA binding domain can activate transcription when simply bound to another protein containing an activation domain; this principle forms the basis for the yeast two-hybrid technique.

In the two-hybrid assay(Fig. 8.3), two fusion proteins are created: the protein of interest (X), which is constructed to have a DNA binding domain attached to its N-terminus, and its potential binding partner (Y), which is fused to an activation domain. If protein X interacts with protein Y, the binding of these two will form an intact and functional transcriptional activator. This newly formed transcriptional activator will then go on to transcribe a reporter gene, which is simply a gene whose protein product can be easily detected and measured. In this way, the amount of the reporter produced can be used as a measure of interaction between our protein of interest and its potential partner.

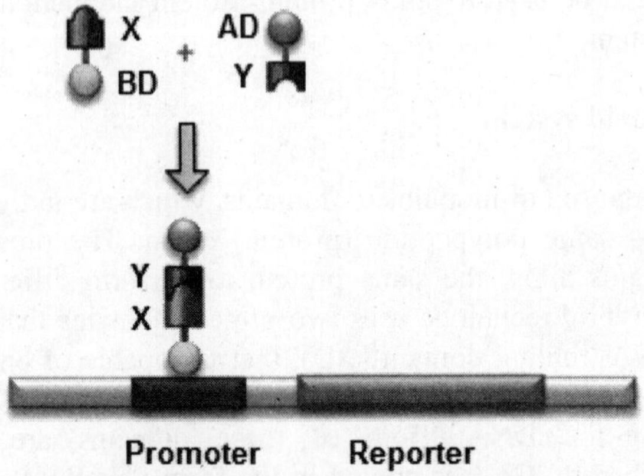

Fig. 8.3: Yeast two hybrid assay

Construction of Vectors

First, it is necessary to construct the 'bait' and 'hunter' fusion proteins. The 'bait' fusion protein is the protein of interest (or 'bait') linked to the GAL4 binding domain, or GAL4 BD. This is done by inserting the segment of DNA encoding the bait into a plasmid, which is a small circular molecule of double-stranded DNA that occurs naturally in both bacteria and yeast. This plasmid will also have inserted in it a segment of Gal4 BD DNA next to the site of bait DNA insertion. Therefore, when the DNA from the plasmid is transcribed and converted to protein, the bait will now have a binding domain attached to its end. The same procedure is used to construct the 'hunter' protein, where the potential binding partner is fused to the GAL4 AD (Figs. 8.4(a) and (b)).

In addition to having the fusion proteins encoded for, these plasmids will also contain selection genes, or genes encoding proteins that contribute to a cell's survival in a particular environment. An example of a selection gene is one encoding antibiotic resistance; when antibiotics are introduced, only cells with the antibiotic resistance gene will survive. Yeast two-hybrid assays typically use selection genes encoding proteins capable of synthesizing amino acids such as histidine, leucine and tryptophan.

Once the plasmids have been constructed, they must next be introduced into a host yeast cell by a process called "transfection". In this process, the outer-membrane of a yeast cell is disturbed by a physical method, such as sonification or chemical disruption. This disruption produces holes that are large enough for the plasmid to enter, and in this way, the plasmids can cross the membrane and enter the cell.

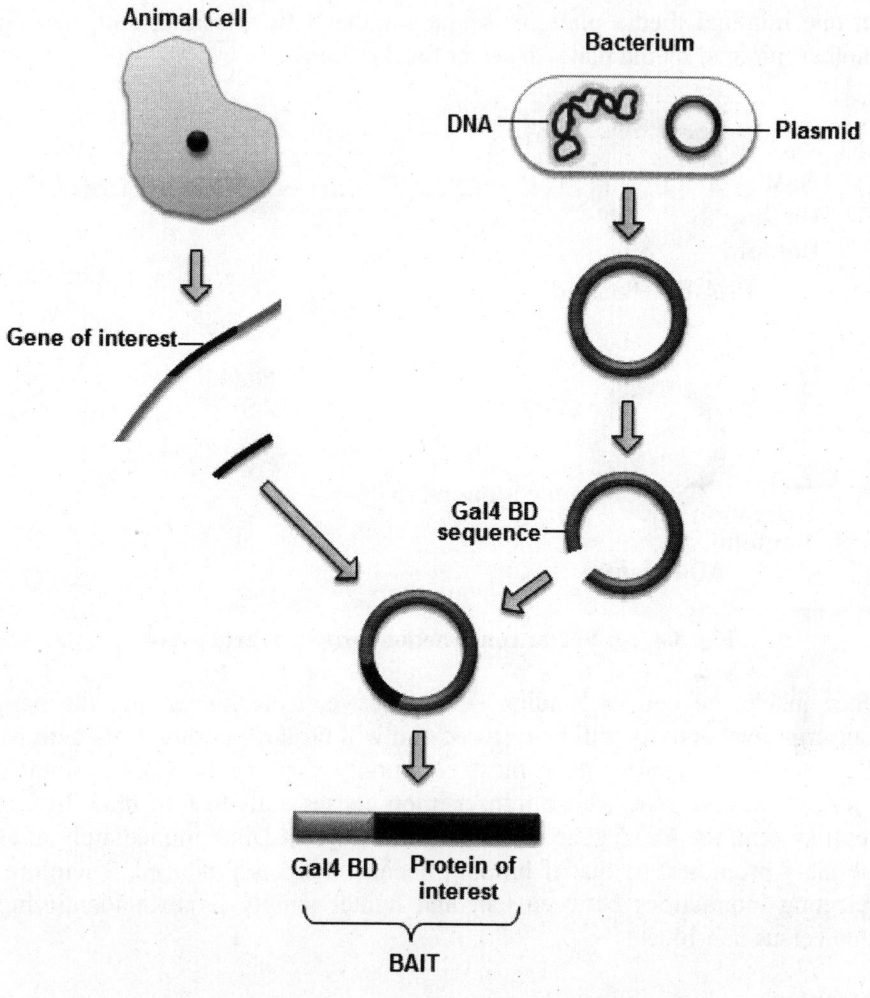

Fig. 8.4 (a): Vector construction for two hybrid assay

Once the cells have been transfected, it is necessary to isolate colonies that have both 'bait' and 'hunter' plasmids. This is because not every cell will have both plasmids cross their plasma membrane; some will have only one plasmid, while others will have none. Isolation of transfected cells involves identifying cells containing plasmids by virtue of their expressing the selection genes mentioned previously. After the cells have been transfected and allowed to recover for several days, they are then plated on minimal media, or media that is lacking one essential nutrient, such as tryptophan. The cells used for transfection are called auxotrophic mutants; these cells are deficient in producing nutrients required for their growth. By supplying the gene for the deficient nutrient in the 'bait' or 'hunter' plasmid, cells containing the plasmid are able to survive on the minimal media, whereas untransfected cells cannot. Selection in this way occurs in two rounds: first on one minimal media plate, to select for the 'bait' plasmid, and then on another minimal media plate, to select for the 'hunter'.

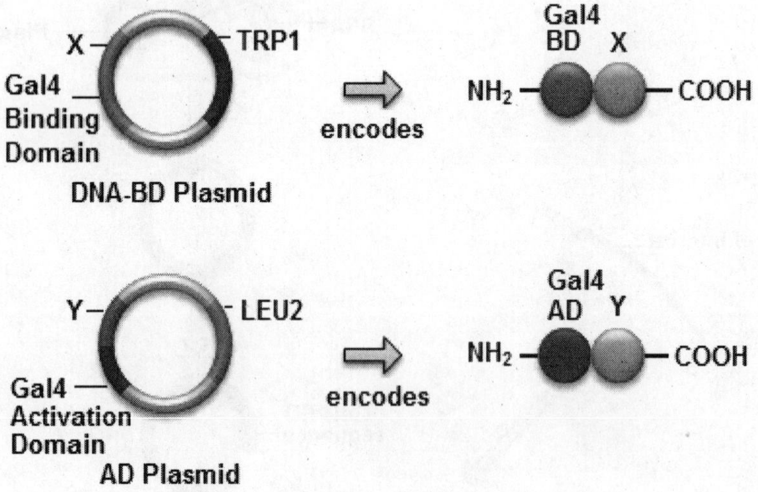

Fig. 8.4 (b): Vector construction for two hybrid assay

Once inside the cell, if binding occurs between the hunter and the bait, transcriptional activity will be restored and will produce normal Gal4 activity (Fig. 8.5). The reporter gene most commonly used in the Gal4 system is LacZ, an *E. coli* gene whose transcription causes cells to turn blue. In this yeast system, the LacZ gene is inserted in the yeast DNA immediately after the Gal4 promoter, so that if binding occurs, LacZ is produced. Therefore, detecting interactions between bait and hunter simply requires identifying blue versus non-blue.

Applications

Generally the yeast two-hybrid assay can identify novel protein-protein interactions. By using a number of different proteins as potential binding partners, it is possible to detect interactions that were previously uncharacterized. Secondly, the yeast two-hybrid assay can be used to characterize interactions already known to occur. Characterization could include determining which protein domains are responsible for the interaction, by using truncated proteins, or under what conditions interactions take place, by altering the intracellular environment.

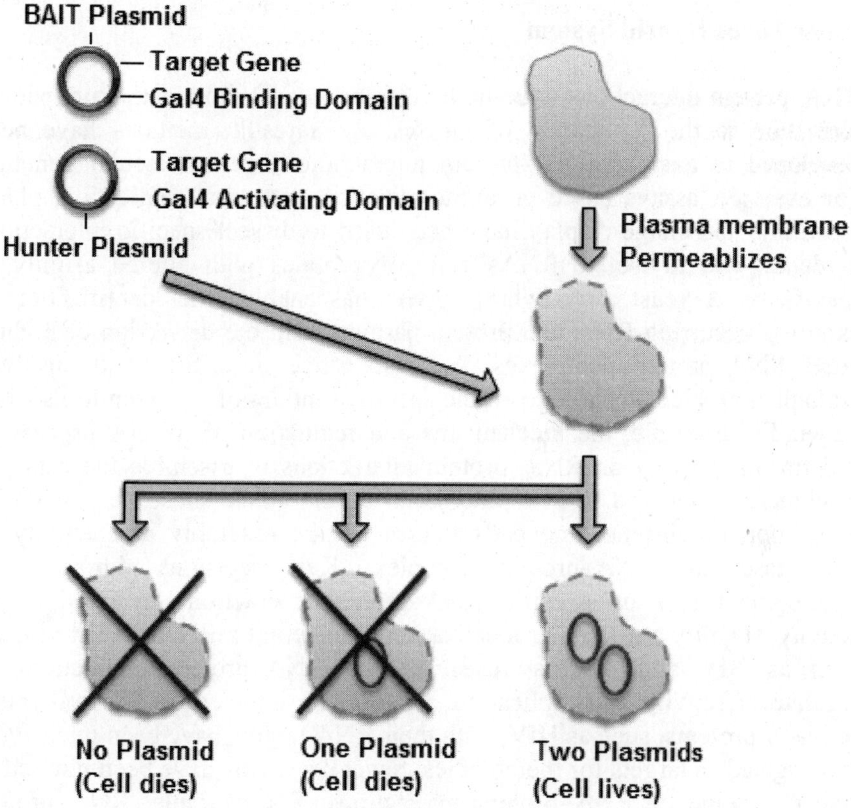

Fig. 8.5: Yeast two hybrid system screening

The last and most recent application of the yeast two-hybrid involves manipulating protein-protein interactions in an attempt to understand its biological relevance. For example, many disorders arise due to mutations causing the protein to be non-functional, or have altered function. Such is the case of some cancers; a mutation in a pro-growth pathway does not allow for the binding of negative regulatory proteins, resulting in the pro-growth pathway never turning 'off'. The yeast two-hybrid is one means of determining how mutation affects a protein's interaction with other proteins. When a mutation is identified that affects binding, the significance of this mutation can be studied further by creating an organism that has this mutation and characterizing its phenotype.

The yeast two-hybrid assay is an elegant means of investigating protein-protein interactions. A fairly new addition to the family of microbiological studies, these interactions have become increasingly important to our understanding of biological systems in the past few years.

Yeast Three Hybrid System

RNA–protein interactions underlie diverse biological processes, from pattern formation to the replication of viruses. As a result, methods have been developed to analyze RNA–protein interactions using molecular genetics. For example, assays based on antitermination, retroviral replication, phage assembly, and phage display have been used to dissect specific interactions in detail, and to isolate RNAs and polypeptides with altered affinity or specificity. A yeast three-hybrid system has enabled the identification of naturally occurring RNA and protein partners, and the dissection of higher-order RNA–protein complexes. A broad range of critical and unsolved biological problems converge on the specific binding of a protein to its RNA target. For example, the mechanisms and regulation of mRNA processing and translation rely on RNA–protein interactions to assemble the catalytic machineries involved and to interact with the RNA substrate. Similarly, replication of chromosome ends hinges on the assembly and activity of telomerase, an RNA–protein complex. Key decisions during early development rely on specific RNA–protein interactions to regulate the activity, stability and cellular localization of maternal mRNAs. RNA viruses, such as HIV and picornaviruses, exploit RNA–protein interactions to regulate infectivity and replication. Indeed, the interactions of viral trans-activator proteins, such as HIV, with their RNA targets have been intensively investigated as targets for therapeutics. Several systems have been devised to detect a range of RNA–protein interactions. For example, RNA–protein interactions can be assayed *in vivo* by placing an RNA binding site in an mRNA such that, when bound to a cognate protein, translation is repressed. Methods based on phage display and the antitermination properties of N protein in bacteria facilitate analysis of interactions between RNA and proteins or peptides and the identification of specificity determinants. *In vitro* selection procedures can be used to identify RNAs that bind with high affinity to a protein of interest and to reveal those features of the RNA that are critical. Yeast three hybrid system is a genetic assay in which specific RNA–protein interactions can be detected rapidly in yeast, in a fashion that is independent of the biological role of the RNA or protein. The approach is based on the yeast two-hybrid system which detects protein–protein interactions. The three-hybrid system presents simple phenotypic properties of yeast, such as the ability to grow or to metabolize a chromogenic compound, to be used to detect and analyze an RNA protein interaction.

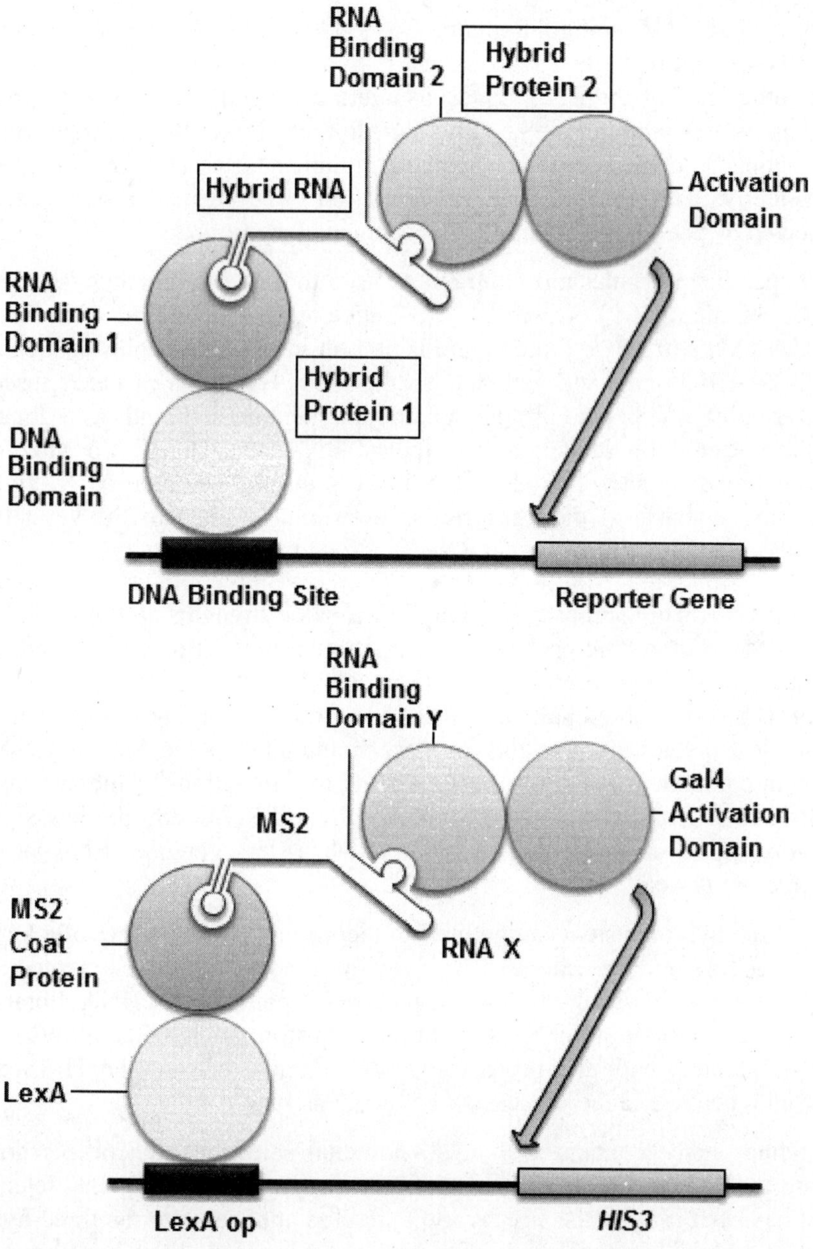

Fig. 8.6: Yeast three hybrid system

Principle of method

The general strategy of the three-hybrid system is same as that of two hybrid system only. DNA binding sites are placed upstream of a reporter gene in the yeast chromosome. A first hybrid protein consists of a DNA binding domain linked to an RNA binding domain. The RNA binding domain interacts with its RNA binding site in a bifunctional (''hybrid'') RNA molecule (Fig. 8.6). The other part of the RNA molecule interacts with a second hybrid protein consisting of another RNA binding domain linked to a transcription activation domain. When this tripartite complex forms at a promoter, even transiently, the reporter gene is turned on. Reporter expression can be detected by phenotype or simple biochemical assays.

The specific molecules most commonly used for three-hybrid analysis are the DNA binding site consists of a 17-nucleotide recognition site for the *Escherichia coli* LexA protein, and is present in multiple copies upstream of both the HIS3 and LacZ genes. Hybrid protein 1 consists of LexA fused to bacteriophage MS2 coat protein, a small polypeptide that binds as a dimer to a short stem–loop sequence. The hybrid RNA consists of two MS2 coat protein binding sites linked to the RNA sequence of interest, X. Hybrid protein 2 consists of the transcription activation domain of the yeast Gal4 transcription factor linked to an RNA binding protein, Y.

The three-hybrid approach has many of the same strengths and limitations of the two-hybrid system to detect protein–protein interactions. By introducing libraries of RNA or protein, cognate partners can be identified. As in two-hybrid screens, the challenge then becomes identifying those that are biologically relevant. For this purpose, mutations in the known RNA or protein component are very useful. Among the limitations of the system are certain technical constraints on producing hybrid RNAs and the low signal-to-noise ratio in screening a library when the genuine RNA–protein interaction is weak.

The three-hybrid system can be used to identify a protein partner of a known RNA sequence. Typically, a three-hybrid reporter strain is created that expresses the RNA of interest as a hybrid molecule. A cDNA library is introduced into this strain by conventional transformation methods. When the RNA interacts with the protein produced from a cDNA, the HIS3 gene becomes active and the yeast grow on media lacking

histidine and/or containing 3-AT. Additional selections can be performed using the LacZ reporter and colorimetric assays. In principle, any selection that has been used in the two-hybrid system is applicable to the three-hybrid system as well. Several successful three hybrid screens have been reported, and have yielded proteins of diverse families of RNA binding proteins.

9

Site Directed Mutagenesis and Protein Engineering

Site specific mutagenesis is the term used to describe when changes in DNA are made at a desired position. The sequence of the bases in DNA literally spells out the information which is used to make a protein. By changing one of these bases, a change in the DNA sequence is produced. This changes the protein that the DNA sequence "spells out".

DNA is a double stranded molecule. One strand will have the code for the protein, and the other strand will be complementary, that is, it will be the "antisense" strand. If these strands are separated, each strand can act as a template and direct the synthesis of its complementary strand.

In site specific mutagenesis, a single strand of DNA, usually 20 to 40 bases long, is synthesized in the lab. This strand of DNA, known as an oligonucleotide or "oligo", will be identical to the gene to be changed, except that the oligo will contain the desired base change (Fig. 9.1).

Using site-directed mutagenesis the information in the genetic material can be changed. A synthetic DNA fragment is used as a tool for changing one particular code word in the DNA molecule. This reprogrammed DNA molecule can direct the synthesis of a protein with an exchanged amino acid. **Michael Smith's** method has become one of biotechnology's most important instruments. **With Smith's site-directed mutagenesis** the researchers can study in detail how proteins function and how they interact with other biological molecules. Site-directed mutagenesis can be used, for example, to systematically change amino acids in enzymes, in order to better understand the function of these important biocatalysts. The researchers can also analyze how a protein is folded into its biologically active three-dimensional structure. The method can also be used to study the complex cellular regulation of the genes and to increase our understanding of the mechanism behind genetic and infectious diseases, including cancer.

METHODS OF *IN VITRO* MUTAGENESIS

Mutagenesis is a fundamentally important DNA technology which seeks to change the base sequence of DNA and test its effect on gene or DNA function. The mutagenesis can be conducted *in vivo* (in studies of model organisms, or cultured cells) or *in vitro* and the mutagenesis can be directed to a specific site in a pre-determined way, or can be random. In the case of *in vivo* mutagenesis, for example, gene targeting offers exquisite site-directed mutagenesis within living cells while exposure of male mice to high levels of a powerful mutagen such as ethyl nitrosurea (ENU) and subsequent mating of the mice offers a form of random mutagenesis which can be important in generating new mutants.

***In vitro* mutagenesis** can involve essentially random approaches to mutagenesis, which may be valuable in producing libraries of new mutants. In addition, if a gene has been cloned and a functional assay of the product is available, it is also very useful to be able to employ a form of *in vitro* mutagenesis which results in alteration of a specific amino acid or small component of the gene product in a predetermined way

Primer extension (the single-primer method) is a simple method for site-directed mutation

The first method of site-directed mutagenesis to be developed was the single-primer method As originally described the method involves *in vitro* DNA synthesis with a chemically synthesized oligonucleotide (7–20 nucleotides long) that carries a base mismatch with the complementary sequence. The method requires that the DNA to be mutated is available in single-stranded form, and cloning the gene in M13-based vectors makes this easy. However, DNA cloned in a plasmid and obtained in duplex form can also be converted to a partially single-stranded molecule that is suitable. The synthetic oligonucleotide primes DNA synthesis and is itself incorporated into the resulting heteroduplex molecule. After transformation of the host *E. coli*, this heteroduplex gives rise to homoduplexes whose sequences are either that of the original wild-type DNA or that containing the mutated base. The frequency with which mutated clones arise, compared with wild-type clones, may be low. In order to pick out mutants, the clones can be screened by nucleic acid hybridization with 32P-labeled oligonucleotide as probe. Under suitable conditions of stringency, i.e. temperature and cation concentration, a positive signal will be obtained only with mutant clones. This allows ready detection of the desired mutant. It is prudent to check the sequence of the mutant directly by DNA sequencing, in order to check that the procedure has not introduced other adventitious changes. This was a particular necessity with early versions of the technique which made use of *E. coli* DNA polymerase. The more recent use of the high-fidelity DNA polymerases has minimized the problem of extraneous mutations as well as shortening the time for copying the second strand. Also, these polymerases do not "strand-displace" the oligomer, a process which would eliminate the original mutant oligonucleotide.

Oligonucleotide mismatch mutagenesis is a popular method of introducing a predetermined single nucleotide change into a cloned gene

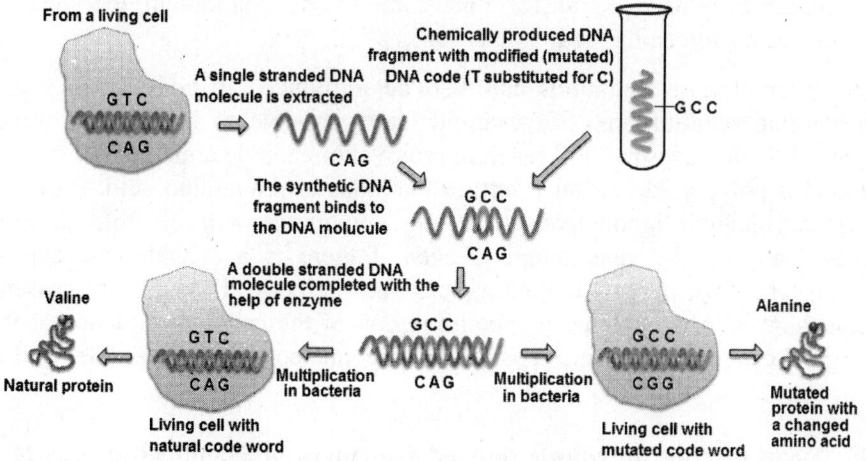

Fig. 9.1: Site Directed mutagenesis

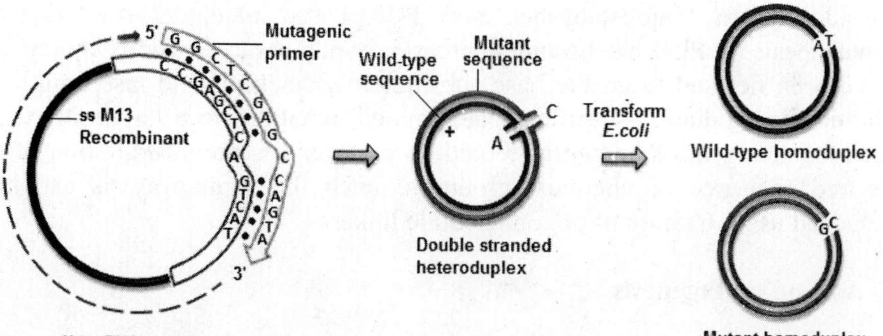

Fig. 9.2: Primer mis match method

Many *in vitro* assays of gene function wish to gain information on the importance of individual amino acids in the encoded polypeptide. This may be relevant when attempting to assess whether a particular missense mutation found in a known disease gene is pathogenic, or just generally in trying to evaluate the contribution of a specific amino acid to the biological function of a protein. A popular general approach involves cloning the gene or cDNA into an M13 (Figs. 9.2 and 9.3) or phagemids vector which permits recovery of single-stranded recombinant DNA. A mutagenic oligonucleotide primer is then designed whose sequence is perfectly complementary to the gene sequence in the region to be mutated, but with a single difference: at the intended mutation site it bears a base that is complementary to the desired mutant nucleotide rather than the original. The mutagenic oligonucleotide is then allowed to prime new DNA synthesis to create a complementary full-length sequence containing the desired mutation. The newly formed heteroduplex is used to transform cells, and the desired mutant genes can be identified by screening for the mutation.

Other small-scale mutations can also be introduced in addition to single nucleotide substitutions. For example, it is possible to introduce a three-nucleotide deletion that will result in removal of a single amino acid from the encoded polypeptide, or an insertion that adds a new amino acid. Provided the mutagenic oligonucleotide is long enough, it will be able to bind specifically to the gene template even if there is a considerable central mismatch. Still larger mutations can be introduced by using cassette mutagenesis in which case a specific region of the original sequence of the original gene is deleted and replaced by oligonucleotide cassettes (Bedwell *et al.*, 1989).

PCR can be used to couple desired sequences or chemical groups to a target sequence and to produce specific pre-determined mutations in DNA sequences

In addition to long-established non PCR based methods, site-directed mutagenesis by PCR has become increasingly popular and various strategies have been devised to enable base substitutions, deletions and insertions. In addition to producing specific predetermined mutations in a target DNA, a form of mutagenesis known as 5′ add-on mutagenesis permits addition of a desired sequence or chemical group in much the same way as can be achieved using ligation of oligonucleotide linkers.

5′ Add-on mutagenesis

This is a commonly used practice in which a new sequence or chemical group is added to the 5′ end of a PCR product by designing primers which have the desired specific sequence for the 3′ part of the primer while the 5′ part of the primer contains the novel sequence or a sequence with an attached chemical group. The extra 5′ sequence does not participate in the first annealing step of the PCR reaction (only the 3′ part of the primer is specific for the target sequence), but it subsequently becomes incorporated into the amplified product, thereby generating a recombinant product. Various popular alternatives for the extra 5′ sequence include:

(i) A suitable restriction site which may facilitate subsequent cell-based DNA cloning.

(ii) A functional component, e.g. a promoter sequence for driving expression, a modified nucleotide containing a reporter group or labeled group, such as a biotinylated nucleotide or fluorophore.

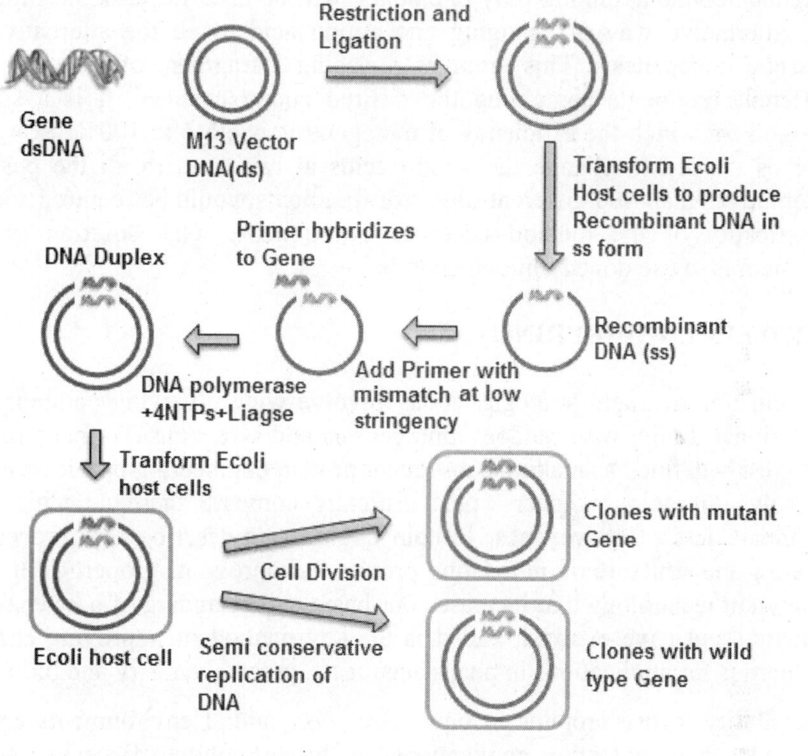

Fig. 9.3: Site directed mutagenesis using M13 vector

Mismatched primer mutagenesis

The primer is designed to be only partially complementary to the target site but in such a way that it will still bind specifically to the target. Inevitably this means that the mutation is introduced close to the extreme end of the PCR product. This approach may be exploited to introduce an artificial diagnostic restriction site that permits screening for a known mutation. Mutations can also be introduced at any point within a chosen sequence using mismatched primers. Two mutagenic reactions are designed in which the two separate PCR products have partially overlapping sequences containing the mutation. The denatured products are combined to generate a larger product with the mutation in a more central location.

Methods have been developed that simplify the process of making all possible amino acid substitutions at a selected site

Using site-directed mutagenesis it is possible to change two or three adjacent nucleotides so that every possible amino acid substitution is made at a site of interest. This generates a requirement for 19 different mutagenic oligonucleotide assuming only one codon will be used for each substitution. An alternative way of changing one amino acid to all the alternatives is cassette mutagenesis. This involves replacing a fragment of the gene with different fragments containing the desired codon changes. It is a simple method for which the efficiency of mutagenesis is close to 100%. However, if it is desired to change the amino acids at two sites to all the possible alternatives than 400 different oligos or fragments would be required and the practicality of the method becomes questionable. One solution to this problem is to use doped oligonucleotide.

PROTEIN ENGINEERING

Protein improvement strategies today involve widely varying combinations of rational design with random mutagenesis and screening. To make further progress—defined as making subsequent protein engineering problems easier to solve—protein engineers must critically compare these strategies and eliminate less effective ones. Protein engineering describes the process of altering the structure of an existing protein to improve its properties. It is an important technology that increases our basic understanding of how enzymes function and have evolved, and it is the key method of improving enzyme properties for applications in pharmaceuticals, green chemistry and biofuels.

The ability to use proteins in unusual or non-natural environments greatly expands their potential applications in biotechnology. Because natural selection has neither maximized the stability of proteins nor optimized them to function under unusual conditions, there is considerable room for their improvement by protein engineering. Significant advances reported within the past year include a dramatic demonstration of a protein's ability to tolerate changes in its amino acid sequence, large increases in protein stability, and the use of random mutagenesis to obtain novel enzymatic properties. Approaches using random or site-directed mutagenesis have been successful in generating proteins able to function in an extended range of environments.

Industrial manufacturing processes often require the use of harsh conditions, specifically toxic chemicals, high temperatures, or extremes in pH. A large portion of biotechnology research is focused on replacing these processes with those mediated by enzymes that can tolerate these extremes and reduce the use of chemicals, or perform the same reactions under more moderate conditions. Success in these areas of research might mean reduced harm to the environment, lower costs in terms of energy used, less dependence on non-renewable resources and less risk to production employees.

Methods for Protein Engineering

A variety of methods have been used and proposed for future use in protein engineering. In this connection mutagenesis, selection, and recombinant DNA are being used and will be increasingly utilized in future.

During the last twenty years, rapid progress has been made in the analysis of protein structure and function. Sequences of amino acids are now available for as many as 8000 proteins, but the three dimensional structures of only about 400 proteins have been resolved through X-ray crystallography.

It is realized that the analysis of protein structure and function has now reached only a level which for DNA existed in 1970's. Such a comparison gets support from the fact that only one centre in the world is available for the storage of protein data (Brookhaven Data Bank), while several such centres are available for DNA sequence data.

From the three dimensional structure of 400 proteins analysed so far, it is apparent that if two proteins are similar in their amino acid sequence, they will tend to fold into similar three dimensional structures. The rules that will translate the amino acid sequence of a protein into its three dimensional structure are sometimes referred to as the second half of the genetic code.

These rules are not fully known yet, but as these rules become known, it will be possible for us to predict three dimensional structure of a protein from its amino acid sequence. This will then also enable us to identify a model structure for a protein and predict its function.

The model can then be used for the synthesis and modification of a gene that will give the desired sequence of amino acid; leading to the three dimensional structure for a specified function. An alternative approach of protein engineering may be to modify a protein by a chemical reaction to make it more suitable for a specific function.

This area of protein research (modification of protein either by recombinant technology or by chemical reaction) is currently receiving considerable emphasis and is popularly described as protein engineering. Through recombinant DNA technology, a gene can now be cloned in an expression vector and made to express in bacteria.

From such bacterial cells having the desired gene cloned in an expression vector, a protein can be obtained in abundant quantity. Since a gene can be artificially synthesized and modified using recombinant DNA technology, this will allow production of novel proteins through genetic engineering, an area popularly described as protein engineering.

The proteins receiving attention in this new and exciting area of protein engineering would include enzymes, synthetic peptides, storage proteins and drugs to be used in medicine, industry and agriculture. The objectives of protein engineering include the following:

(i) To create superior enzymes to catalyze production of high value specific chemicals.

(ii) To produce enzymes for large scale use in the chemical industry.

(iii) To produce biological compounds (including synthetic peptides, storage proteins and specific drugs), that are superior to natural ones. Some success in this area of protein engineering has already been achieved and much progress is expected to be made in this area in future.

Rationale of Protein Enzyme Engineering

Although thousands of proteins have been characterized in prokaryotes and eukaryotes, only few became commercially important. This is due to the high cost of isolating and purifying enzymes in sufficient quantities.

Although the cost aspect has been overcome by producing an enzyme in large quantities in bacteria, for its industrial application, an enzyme (outside the cell) should also have some characteristics in addition to those of enzymes in the cells. These characteristics may include the following:

(i) Enzyme should be robust with a long life.

(ii) Enzyme should be able to use the substrate supplied in the industry even if it differs slightly from that in the cell.

(iii) Enzyme should be able to work under conditions (e.g. extremes of pH, temperature and concentration) of the industry even if they differ from those in the cell.

In view of the above, enzyme should be engineered to meet the altered needs. Therefore, efforts have been made to alter the properties of the enzymes. Following is the list of properties that one needs to alter in a predictable manner for protein or enzyme engineering.

(1) Kinetic properties of enzyme turnover and Michaelis Constant, Km.

(2) Theremostability and the optimum temperature for the enzyme.

(3) Stability and activity of enzyme in nonaqueous solvents.

(4) Substrate and reaction specificity.

(5) Cofactor requirements.

(6) Optimum pH.

(7) Protease resistance.

(8) Allosteric regulation.

(9) Molecular weight and subunit structure.

For a particular class of enzymes, variation in nature may occur for each of the above properties, so that one may like to combine the optimum properties to get the most efficient form of the enzyme.

This aspect of protein engineering will be illustrated using the example of glucose isomerases, which convert glucose into other isomers like fructose and are used to make high fructose corn syrup vital for soft drink industry. It exhibits wide variation in its properties.

Sometimes, it may not be possible to get a combination of optimum properties. For instance, an enzyme with highest activity may not be the most stable. Therefore, a compromise in properties may have to be made, if we have to select an enzyme from the available variability or even if we create variability by mutagenesis.

However, if structure and function relationship of an enzyme is known, the structural features for desirable function may be combined and protein engineering techniques may then be used to create a novel enzyme exhibiting a combination of all desirable functional properties.

Glucose isomerase belongs to a TIM barrel family of enzymes which resemble each other in having a highly characteristic domain called TIM barrel, with active site for catalytic action at one end. This TIM barrel may be found in enzymes that may differ in sequences and may catalyze different reactions.

As earlier discussed, since similarities of structure of protein meant similarities in function, TIM barrel presents a challenge to this concept. However, it is curious that some enzymes in this family appear in pairs in their metabolic pathways so that they catalyze two consecutive steps thus showing coupling of their functions. As an example of two enzymes of TIM barrel family, while 'triose phosphate isomerase' is one of the most efficient catalysts, 'glucose isomerase' is relatively very inefficient.

Therefore, if 'glucose isomerase' enzyme is redesigned to use the highly efficient domain of TIM barrel family, it will be a remarkable achievement for soft drink industry. Efforts in this direction are being made.

Assumptions for Protein Engineering

While attempting protein engineering, one should recognize the following properties of enzymes:

(i) Many amino acid substitutions, deletions or additions lead to no change in enzyme activity, so that they are silent mutations.

(ii) Proteins have a limited number of basic structures and only minor changes are superimposed on them leading to variation;

(iii) Similar patterns of chain folding and domain structure can arise from different amino acid sequences, which show little or no homology (although same amino acid sequence never gives different folding and domain structures).

The above properties suggest that while many major changes sometimes may lead to no alteration in function, some of the minor changes at specific positions may lead to the desired favorable change. For example, a single amino acid replacement (glycine to aspartic acid) in *E. coli* asparate transcarbamylase leads to:

(i) Loss of activity.

(ii) An alteration in the binding of catalytic and regulatory subunits.

Another example involved the engineering of a single chain biosynthetic antibody binding site (BARS), which is though only one sixth of the size of the complete antibody, but retains its antigen binding specificity.

This synthetic fragment has heavy and light chain variable regions (V H and V J connected by a 15 - amino acid linker. A synthetic gene has also been prepared for the fragment, which expressed in *E. coli*. This fragment binds to digoxin, a cardiac glycoside. Single amino acid replacements in BABS fragment have sometimes led to major changes in its binding affinity. In view of the above, it is necessary to examine not only the crystal structure but also the active sites therein, so that the gene may be modified or artificially synthesized for protein engineering to meet the desired needs.

Mutagenesis and Selection for Protein Engineering

Mutagenesis and selection can be effectively utilized for improving a specific property of an enzyme. Following are some of the examples of selection of mutant enzymes:

(i) *E. coli* anthranilate synthetase enzyme is normally sensitive to tryptophan inhibition due to feedback inhibition. An MTR 2 mutation of *E. coli* was found to possess an altered form of enzyme anthranilate synthetase that is insensitive to tryptophan inhibition.

They may help in continuous synthesis of tryptophan without any inhibition by tryptophan accumulated as a product.

(ii) Xanthine dehydrogenase enzyme oxidizes 2 hydroxy-purine at position 8, but a mutant has been inolated which oxidizes 2 hydroxy-purine at position 6.

(iii) Lactate dehydrogenase (LDU) from a bacterial system was modified to malate dehydrogenase able a natural mutation leading to a single amino acid substitution (Gln to Arg).

In the above and other cases of naturally occurring mutant enzymes, single amino acid modification or addition/deletion has been observed.

However, if improvement requires changes in several amino acids, such a mutant will be rare or nonexistent and modifications of this type will be possible only through gene modification techniques discussed in the following section.

Site-directed mutagenesis is a molecular method made possible by modern biochemical techniques such as PCR and gene sequencing. It involves doing exactly what it sounds like – creating a genetic mutation at a specific site in a gene. Of all the means of altering protein structure, this is the method that requires the most detailed knowledge about the gene of interest. The complete nucleotide sequence needs to be known, or at least enough to generate primers for PCR at either end, and in the region that is to be mutated.

Generally the active site of a protein is mutated (the site of substrate binding and reaction catalysis), in order to enhance substrate binding, or otherwise improve rates of reaction. Other regions that might have to do with enzyme activity and warrant attention might be those having to do with protein folding (eg. formation of disulfide bonds between cystine residues), cofactor binding, or charge (eg. acid-base catalysis) and ligand (eg. carboxy, acetyl, amino, phosphate, hydroxyl group) interactions.

To perform site-directed mutagenesis, the gene of interest is cloned by PCR, but using primers in the region of interest that introduce a single site (base-pair) mutation resulting in a codon for a different amino acid at that site on the resulting protein. Generally the site of interest will be somewhere in the middle of the gene, so the gene is cloned in 2 or more fragments, with restriction enzyme sites introduced by the primers that allow for the resulting fragments to be ligated (connected) together in the end. Another form of site-directed mutagenesis is to completely remove a section of the gene and ligating the remaining pieces to generate a fusion protein.

Chemical modification and protein engineering especially are now the useful tools for thermo-stabilizing proteins, and also for elucidating the mechanism of protein stability. The information on the mechanism so far accumulated indicate that a single or few amino acid replacement(s) in a protein is/are sufficient to enhance protein thermostability. Salt bridges inside protein molecule or decrease of internal or external hydrophobicity, respectively, may contribute to increased thermostability. However, generalized molecular reasons for protein thermostability and generalized methods for protein stabilization have not yet been proposed. Some of typical examples of the application of protein engineering to stabilize proteins are presented. They are based on information concerning the tertiary structure of the proteins or their related proteins. Even if such structural information is unavailable, one can replace amino acid(s) in a protein by mutagenesis of the gene coding for the protein via the application of chemicals to the gene (or the plasmid harbouring the gene) or organism.

10

Techniques Associated with cloning

Fuelled by the drive to complete the Human Genome Project, many laboratories have developed new methods of screening clone libraries. From PCR-based strategies to pooling schemes and increased automation, the tedious task of library screening has become less labor-intensive and more cost-efficient. Currently, two main screening methods dominate: hybridization and polymerase chain reaction (PCR). Hybridization techniques include the Southern, Northern and Western Blotting techniques and besides them other techniques like Primer Extension, Reporter assay, RNase Protection assay are also used to analyze the clones. These techniques are valuable tools of the Recombinant DNA technology.

BLOTTING TECHNIQUES

Blotting is the technique in which nucleic acids or proteins are immobilized onto a solid support generally nylon or nitrocellulose membranes. Blotting of nucleic acid is the central technique for hybridization studies. Nucleic acid labelling and hybridization on membranes have formed the basis for a range of experimental techniques involving understanding of gene expression, organization, etc. When genomic DNA, extracted from any tissue of a plant or animal species, is digested with a restriction enzyme, it is cleaved into segments. The segments of different sizes can be separated through gel electrophoresis before a molecular probe is used to detect the segments which have sequences similar to those in the probe. Gel electrophoresis involves movement of fragments or molecules under a high voltage electric current. The mixture of DNA fragments is loaded in a well created on one edge of the gel. The gel may be a cylinder or a slab (usually a slab for cloning expt.), about 10 cm long and 0. 5 cm thick. The rate of movement of fragments is inversely correlated with the size of fragments or molecules, so that heavier fragments will remain closer to the site of loading and the lighter fragments will move away.

Fragments of different sizes will appear as bands on the gel and can be examined or isolated for further study. More often agarose gels are used, but for separation of fragments differing by few base pairs, polyacrylamide gels are used. Polyacrylamide gels are more commonly used for DNA sequencing experiments. The technique of gel electrophoresis is used for separation of DNA molecules of different sizes. However, DNA molecules of large size could not be handled in this technique. In recent years, a new technique called pulsed field gel electrophoresis (PFGE) has been used, for separation of large sized DNA molecules, sometimes representing whole chromosomes.

Using this technique, separation of DNA molecules belonging to each of the 16 individual chromosomes of yeast has become possible. In this technique, short pulses of electricity are used in two different directions and DNA is embedded and used in the form of agarose plugs to avoid fragmentation of large DNA molecules.

Using the technique of PFGE and a more refined technique CHEFE (countour clamped homogeneous electric field electrophoresis), genome of several fungi could be resolved into chromosomal bands and used for mapping of DNA sequences on specific chromosomes.

A mixture of DNA, RNA or protein fragments can be separated by gel electrophoresis and the separated bands can be stained and visualized directly in the gel. However, to confirm the identity of these bands or to find similarity of one or more of these bands with a known and available molecular probe, it is possible to hybridize these bands with a labeled probe. To facilitate this hybridization, the bands are often transferred to a nitrocellulose membrane through a technique described as blotting. When DNA bands are thus blotted, this is called Southern blotting (after the name of' E.M. Southern); when RNA bands are thus transferred it is described by the jargon term Northern blotting and similarly when protein bands are transferred, the technique is describe as Western blotting.

Southern Blotting

Southern blotting is a technique named after its inventor and developer, the British biologist Edwin M. Southern in 1975. Southern blotting is a technique which allows the detection of a specific DNA sequence (gene or other) in a large, complex sample of DNA (e.g. cellular DNA). As mentioned, the Southern blot is a technique used to identify and locate DNA sequences which are complementary to another piece of DNA called a probe.

The separation on an electrophoretic gel of sequences or fragments of genomic or complementary DNA, partially digested by endonucleases, The fragments are then 'blotted' onto a membrane and allowed to hybridize with a specific, labeled probe in order to detect which bands contain the fragment or gene of interest. Southern blot is particularly useful in detecting large gene rearrangements/deletions and large trinucleotide repeat expansions.

A Southern blot is a method routinely used in molecular biology to check for the presence of a DNA sequence in a DNA sample. Southern blotting combines agarose gel electrophoresis for size separation of DNA with methods to transfer the size-separated DNA to a filter membrane for probe hybridization. Other blotting methods (i.e., western blot, northern blot, southwestern blot) that employ similar principles, but using RNA or protein, have later been named in reference to Southern's name. As the technique was eponymously named, Southern blot should be capitalised, whereas northern and western blots should not. It is also used to determine the molecular weight of a restriction fragment and to measure relative amounts in different samples. Blots are techniques for transferring DNA and RNA proteins onto a carrier so they can be separated, and often follows the use of a gel electrophoresis. The Southern blot is used for transferring DNA. Hybridization of the probe to a specific DNA fragment on the filter membrane indicates that this fragment contains DNA sequence that is complementary to the probe.

For Southern blotting, DNA sample is first digested with a restriction enzyme and digested sample is gel electrophoresed. The DNA bands in the gel are denatured into single strands with the help of an alkali solution.

Subsequently, the gel is laid on top of a buffer saturated filter paper, placed on a solid support (e.g. glass plate), with its two edges immersed in the buffer. A sheet of nitrocellulose membrane is placed on top of the gel and a stack of many papers (paper towels) on top of this membrane.

A weight of about 0.5 kg is placed on top of paper towels. The buffer solution is drawn up by filter paper wick, and passes through the gel to the nitrocellulose membrane and finally to the paper towels. While passing through the gel, the buffer carries with it single stranded DNA, which binds on to the nitrocellulose membrane, when the buffer passes through it to the paper towels (Fig. 10.1).

After leaving this arrangement for a few hours or overnight, paper towels are removed and discarded. The nitrocellulose membrane with single stranded DNA bands blotted on to it, is baked at 80.C for 2-3 hours to fix the DNA permanently on the membrane.

This membrane now has a replica of DNA bands from Agarose gel, and can be used for hybridization with radioactively labeled DNA or RNA probe. The membrane may then be washed to remove any unbound DNA and X-ray film is exposed to the hybridized membrane to get autoradiograph.

Applications of the Southern Blot Method

Southern blots are used in several main areas including gene discovery and mapping, evolution and development studies, diagnostics and forensics.

- Southern blots allow investigators to determine the molecular weight of a restriction fragment and to measure relative amounts in different samples.
- Southern blot is used to detect the presence of a particular bit of DNA in a sample.
- Analyze the genetic patterns which appear in a person's DNA.
- Analyze restriction digestion fragmentation of DNA or a biological sample

In regards to genetically modified organisms, Southern blotting is used as a definitive test to ensure that a particular section of DNA of known genetic sequence has been successfully incorporated into the genome of the host organism.

Northern Blotting

Initially the technique of Southern blotting used for DNA transfer from gel to the membrane could not be used for blot-transfer of RNA. Instead mRNA bands from the gel were blot-transferred onto a chemically reactive paper, prepared by diazotization of aminobenzyloxymethyl paper.

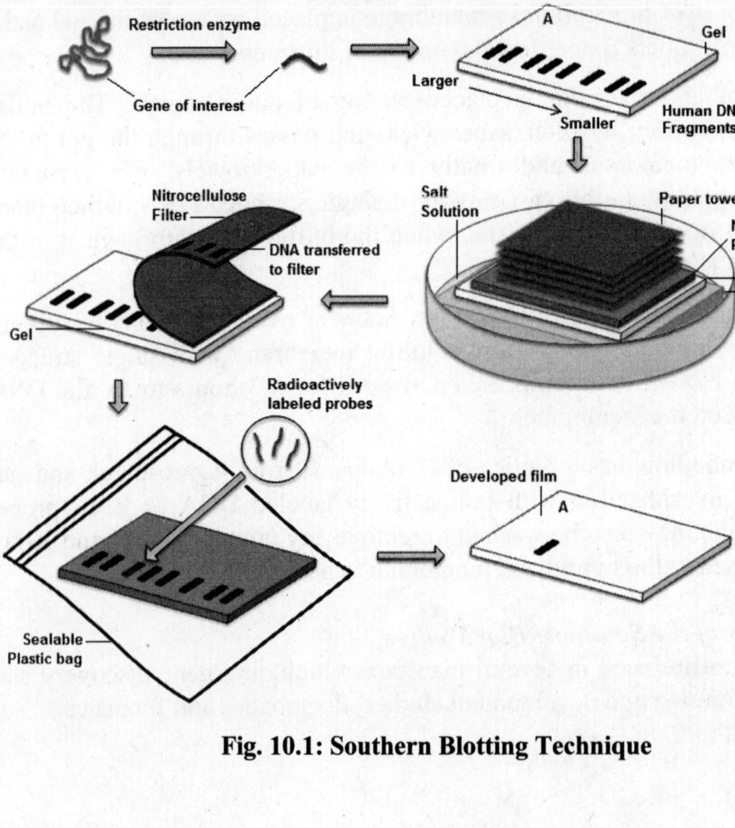

Fig. 10.1: Southern Blotting Technique

The technique being related to Southern blotting was called Northern blotting (not after the name of any scientist as in Southern blotting). Later, it was shown that mRNA bands can be hybridized with a labelled DNA or RNA probe. The single stranded regions of the hybridized probe are removed by nuclease (mungbean nuclease or S-l nuclease), so that quantitative estimation of hybridized mRNA can also be made.

The **northern blot** is a technique used in molecular biology research to study gene expression by detection of RNA (or isolated mRNA) in a sample. With northern blotting it is possible to observe cellular control over structure and function by determining the particular gene expression levels during differentiation, morphogenesis, as well as abnormal or diseased conditions. Northern blotting involves the use of electrophoresis to separate RNA samples by size and detection with a hybridization probe complementary to part of or the entire target sequence. The term 'northern blot' actually refers specifically to the capillary transfer of RNA from the electrophoresis gel to the blotting membrane, however the entire process is commonly referred to as northern blotting. The northern blot technique was developed in 1977 by James Alwine, David Kemp, and George Stark at Stanford University. Northern blotting takes its name from its similarity to the first blotting technique, the Southern blot, named for biologist Edwin Southern. The major difference is that RNA, rather than DNA, is analyzed in the northern blot.

The Northern blotting procedure is straightforward and provides opportunities to evaluate progress at various points (e.g., integrity of the RNA sample and how efficiently it has transferred to the membrane). RNA samples are first separated by size via electrophoresis in an agarose gel under denaturing conditions (Fig. 10.2). The RNA is then transferred to a membrane, crosslinked and hybridized with a labeled probe. Northern hybridization is exceptionally versatile in that radiolabeled or nonisotopically labeled DNA, in vitro transcribed RNA and oligonucleotides can all be used as hybridization probes. Additionally, sequences with only partial homology (e.g., cDNA from a different species or genomic DNA fragments that might contain an intron) may be used as probes.

The steps involved in Northern analysis include:

- RNA isolation (total or poly(A) RNA)
- Probe generation
- Denaturing Agarose gel electrophoresis
- Transfer to solid support and immobilization
- Prehybridization and hybridization with probe
- Washing

- Detection
- Stripping and reprobing (optional)

Once separated by denaturing agarose gel electrophoresis, the RNA is transferred to a positively charged nylon membrane and then immobilized for subsequent hybridization. The best low-tech method for agarose transfer is by a passive, slightly alkaline, downward elution. This procedure, in comparison to upward transfer, is much faster and therefore results in tighter bands and more signal. Alternatively, commercially available active transfer methods (electroblotter, semidry electroblotter, vacuum blotter, pressure blotter, etc.) can be used. Prehybridization, or blocking, is required prior to probe hybridization to prevent the probe from coating the membrane. Good blocking is necessary to minimize background problems. After hybridization, unhybridized probe is removed by washing in several changes of buffer. If a radio labeled probe was used, the blot can be wrapped in plastic wrap to keep it from drying out and then immediately exposed to film for autoradiography. If a nonisotopic probe was used, the blot must be treated with nonisotopic detection reagents prior to film exposure.

Despite these advantages, there are limitations associated with Northern analysis.

First, if RNA samples are even slightly degraded, the quality of the data and the ability to quantitate expression are severely compromised. For example, even a single cleavage in 20% of 4 kb target molecules will decrease the returned signal by 20%. Thus, Rnase-free reagents and techniques are essential.

Second, a standard Northern procedure is, in general, less sensitive than nuclease protection assays and RT-PCR, although improvements in sensitivity can be achieved by using high specific activity antisense RNA probes, optimized hybridization buffers and positively charged nylon membranes. Sensitivity can be further improved with oligo dT selection for enrichment of mRNA, since physical constraints of gel electrophoresis and membrane transfer limit the amount of RNA that can be analyzed without loss of resolution and saturation of the transfer membrane.

A third limitation of Northern blotting has been the difficulty associated with multiple probe analysis. To detect more than one message, it is usually necessary to strip the initial probe before hybridizing with a second probe. This process can be time consuming and problematic, since harsh treatment is required to strip conventional probes from blots.

Northern analysis remains a standard method for detection and quantitation of mRNA levels despite the advent of powerful techniques, such as RT-PCR, gene array analysis and nuclease protection assays. Northern analysis provides a direct relative comparison of message abundance between samples on a single membrane. It is the preferred method for determining transcript size and for detecting alternatively spliced transcripts.

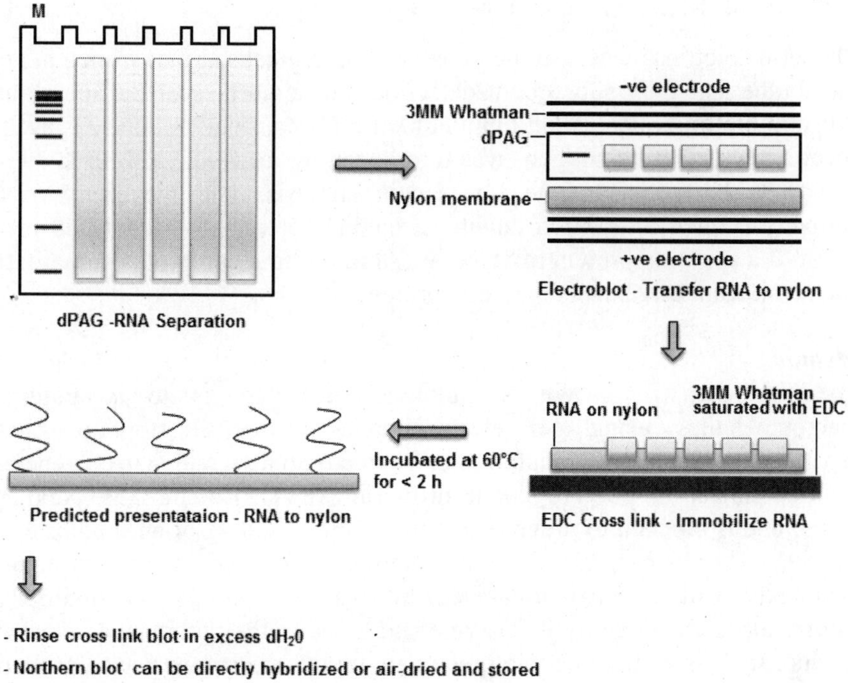

Fig. 10.2: Northern Blotting

Western Blotting

Western blotting is a technique used to identify and locate proteins based on their ability to bind to specific antibodies. Western blot analysis can detect your protein of interest from a mixture of a great number of proteins. Western blotting can give you information about the size of your protein (with comparison to a size marker or ladder in kDa), and also give you information on protein expression (with comparison to a control such as untreated sample or another cell type or tissue). Western blotting, proteins are transferred from the polyacrylamide gel (PAGE) or sodium dodecyl sulphate PAGE onto the nitrocellulose membrane or nylon membrane. In this method there is no requirement for pre-treatment as the proteins are small and in most cases they are not linear. The second difference is that there is no need for cross linking of the proteins to the membrane.

After the transfer of proteins from the gel onto the membrane, it is incubated with or in a solution containing antibodies against protein of interest. Non-bound antibodies are washed off the membrane and the presence of the initial antibody is detected by placing the membrane in a solution containing a secondary antibody. These secondary antibodies react with immunoglobulins or primary antibodies. This secondary antibody is conjugated to either a radioactive isotope or an enzyme that produces visible colour to analyse the protein expression and regulation.

The term "blotting" refers to the transfer of biological samples from a gel to a membrane and their subsequent detection on the surface of the membrane. Western blotting (also called immunoblotting because an antibody is used to specifically detect its antigen) was introduced by Towbin, et al. in 1979 and is now a routine technique for protein analysis. The specificity of the antibody-antigen interaction enables a target protein to be identified in the midst of a complex protein mixture. Western blotting can produce qualitative and semiquantitative data about that protein.

Method

The first step in a Western blotting procedure is to separate the macromolecules using gel electrophoresis. After electrophoresis, the separated molecules are transferred or blotted onto a second matrix, generally a nitrocellulose or polyvinylidene difluoride (PVDF) membrane. Next, the membrane is blocked to prevent any nonspecific binding of antibodies to the surface of the membrane. Most commonly, the transferred protein is complexed with an enzyme-labeled antibody as a probe. An appropriate substrate is then added to the enzyme and together they produce a detectable product such as a chromogenic precipitate on the membrane for colorimetric detection. The most sensitive detection methods use a chemiluminescent substrate that, when combined with the enzyme, produces light as a byproduct. The light output can be captured using film, a CCD camera or a phosphorimager that is designed for chemiluminescent detection. Alternatively, fluorescently tagged antibodies can be used, which are directly detected with the aid of a fluorescence imaging system. Whatever system is used, the intensity of the signal should correlate with the abundance of the antigen on the membrane.

Detailed procedures for detection of a Western blot vary widely. One common variation involves direct vs. indirect detection. With the direct detection method (Fig. 10.3), the primary antibody that is used to detect an antigen on the blot is labeled with an enzyme or fluorescent dye. This detection method is not widely used as most researchers prefer the indirect detection method for a variety of reasons. In the indirect detection method (Fig. 10.4), a primary antibody is added first to bind to the antigen. This is followed by a labeled secondary antibody that is directed against the primary antibody. Labels include biotin, fluorescent probes such as fluorescein or rhodamine, and enzyme conjugates such as horseradish peroxidase or alkaline phosphatase. The indirect method offers many advantages over the direct method.

Direct Method

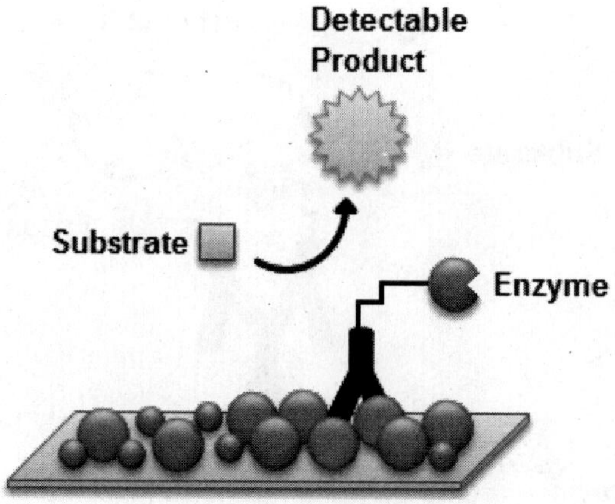

Fig. 10.3: Direct detection method

Advantages

- Quicker since only one antibody is used.
- No concern for cross-reactivity of a secondary antibody.
- Double possible with different labels on primary antibodies.

Disadvantages

- Labeling may reduce immunoreactivity of primary antibody.
- Labeled primary antibodies are expensive.
- Low flexibility in choice of primary antibody label.
- Little signal amplification.

Indirect Method

Advantages

- Secondary antibody can amplify signal
- A variety of labeled secondary antibodies are available
- One secondary may be used with many primary antibodies
- Labeling does not affect primary antibody immunoreactivity
- Changing secondary allows change of detection method

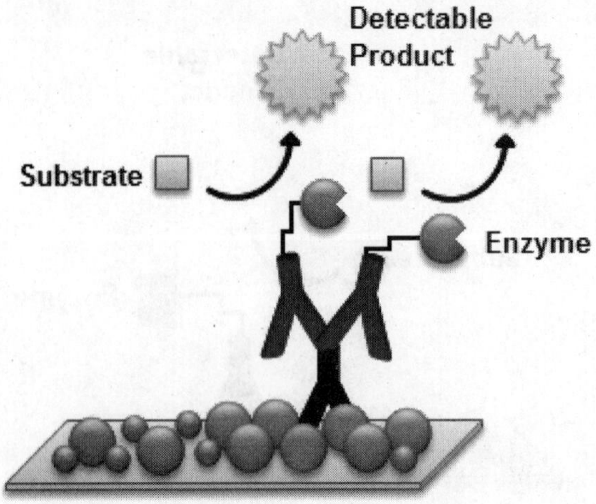

Fig. 10.4: Indirect detection method

Disadvantages

- Secondary antibodies may produce nonspecific staining
- Additional steps required compared to the direct method

Electrophoretic Separation of Proteins

Gel electrophoresis is a technique in which charged molecules, such as protein or DNA, are separated according to physical properties as they are forced through a gel by an electrical current. Proteins are commonly separated using polyacrylamide gel electrophoresis (PAGE) to characterize individual proteins in a complex sample or to examine multiple proteins within a single sample. When combined with Western blotting, PAGE is a powerful analytical tool providing information on the mass, charge, purity or presence of a protein. Several forms of PAGE exist and can provide different types of information about the protein(s).

Transfer Proteins to a Membrane

Following electrophoresis, the protein must be transferred from the electrophoresis gel to a membrane. There are a variety of methods that have been used for this process, including diffusion transfer, capillary transfer, heat-accelerated convectional transfer, vacuum blotting transfer and electro elution. The transfer method that is most commonly used for proteins is electro elution or electrophoretic transfer because of its speed and transfer efficiency. This method uses the electrophoretic mobility of proteins to transfer them from the gel to the membrane. Electrophoretic transfer of proteins involves placing a protein-containing polyacrylamide gel in direct contact with a piece of nitrocellulose or other suitable, protein-binding support and "sandwiching" this between two electrodes submerged in a conducting solution. When an electric field is applied, the proteins move out of the polyacrylamide gel and onto the surface of the membrane, where the proteins become tightly attached. The result is a membrane with a copy of the protein pattern that was originally in the polyacrylamide gel(Fig. 10.5).

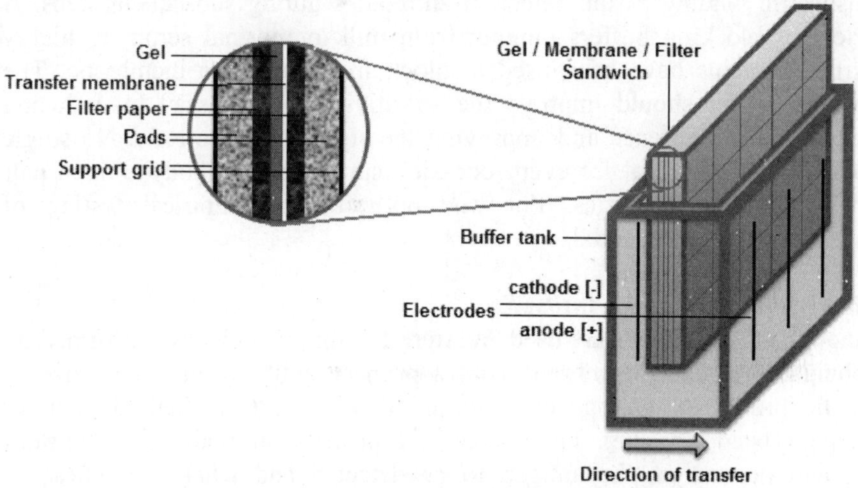

Fig. 10.5: Transfer of proteins to membrane

Transfer efficiency can vary dramatically among proteins, based upon the ability of a protein to migrate out of the gel and its propensity to bind to the membrane under a particular set of conditions. The efficiency of transfer depends on factors such as the composition of the gel, complete contact of the gel with the membrane, the position of the electrodes, the transfer time, size and composition of proteins, field strength and the presence of detergents and alcohol in the buffer. Optimal transfer of proteins is generally obtained in low ionic strength buffers and with low electrical current. After transfer and before proceeding with the Western blot, it is often desirable to stain all proteins on the membrane with a reversible stain to check the transfer efficiency. Although the gel may be stained to determine that protein has been moved out of the gel, this does not ensure efficient binding of protein on the membrane so it is desirable to stain the proteins on the membrane itself. Because dyes may interfere with antibody binding and detection, a protein stain that is easily removed is required. Ponceau S stain is the most widely used reagent for reversibly staining proteins on a membrane. The Ponceau S membrane stain has limited sensitivity, does not photograph well and fades quickly which makes documentation difficult. Superior alternatives for staining protein on nitrocellulose or PVDF membranes are available which allow the detection of low nanogram levels of protein, are easily photographed and do not fade until removed.

Blocking Nonspecific Sites

The membrane supports used in Western blotting have a high affinity for proteins. Therefore, after the transfer of the proteins from the gel, it is important to block the remaining surface of the membrane to prevent nonspecific binding of the detection antibodies during subsequent steps. A variety of blocking buffers ranging from milk or normal serum to highly purified proteins have been used to block free sites on a membrane. The blocking buffer should improve the sensitivity of the assay by reducing background interference and improving the signal to noise ratio. No single blocking agent is ideal for every occasion since each antibody-antigen pair has unique characteristics. For true optimization, empirical testing of blocking buffers is essential.

Primary and Secondary Antibodies

Although other methods are used, Western blotting is typically performed by probing the blocked membrane with a primary antibody that recognizes a specific protein or epitope on a group of proteins (i.e., SH2 domain or phosphorylated tyrosine). The choice of a primary antibody for a Western blot will depend on the antigen to be detected and what antibodies are available to that antigen. It is also important to note that not all primary antibodies are suitable for Western blotting and the feature should be verified, if possible, before purchasing a new primary antibody.

In general, the primary antibody which recognizes the target protein in a Western blot is not directly detectable. Therefore, tagged secondary antibodies or other detection reagents are used as the means of ultimately detecting the target antigen (indirect detection). A wide variety of labeled secondary detection reagents can be used for Western blot detection. The choice of which depends upon either the species of animal in which the primary antibody was raised (the host species) or any tag on that antibody (i.e., biotin or DIG). For example, if the primary antibody is an unmodified mouse monoclonal antibody then the secondary antibody must be an anti-mouse IgG secondary antibody obtained from a non-mouse host.

Antibody dilutions are typically made in the wash buffer. The presence of detergent and a small amount of the blocking agent in the antibody diluent often helps to minimize background thereby increasing the signal:noise ratio. Conversely, adding too much blocking agent or detergent to the antibody dilution solution can prevent efficient binding of the antibody to the antigen, causing reduced signal as well as reduced background.

Detection Methods

While there are many different tags that can be conjugated to a secondary or primary antibody, the detection method used will limit the choice of what can be used in a Western blotting assay. Radioisotopes were used extensively in the past, but they are expensive, have a short shelf-life, offer no improvement in signal: noise ratio and require special handling and disposal. Alternative labels are enzymes and fluorophores.

Enzymatic labels are most commonly used for Western blotting and, although they require extra steps, can be extremely sensitive when optimized with an appropriate substrate. Alkaline phosphatase (AP) and horseradish peroxidase (HRP) are the two enzymes used most extensively as labels for protein detection. An array of chromogenic, fluorogenic and chemiluminescent substrates are available for use with either enzyme. Alkaline phosphatase offers a distinct advantage over other enzymes in that its reaction rate remains linear allowing sensitivity to be improved by simply allowing a reaction to proceed for a longer time period. Unfortunately, the increased reaction time often leads to high background signal resulting in low signal: noise ratios. Horseradish peroxidase (HRP) conjugated antibodies are considered superior to antibody-AP conjugates with respect to the specific activities of both the enzyme and antibody due the smaller size of HRP enzyme and compatibility with conjugation reactions. In addition, the high activity rate, good stability, low cost and wide availability of substrates makes HRP the enzyme of choice for most applications.

Enzyme conjugated antibodies offer the most flexibility in detection and documentation methods for Western blotting because of the variety of substrates available. The simplest detection/documentation system is to use chromogenic substrates. While not as sensitive as other substrates, the chromogenic substrates allow direct visualization of blot development. Unfortunately, chromogenic substrates tend to fade as the blot dries or when stored making the blot itself an unreliable means of documentation. However, it is fairly straightforward to either photocopy or directly scan the blot in order to make a permanent replica of chromogenic Western blot results.

Chemiluminescent blotting substrates differ from other substrates in that the signal is a transient product of the enzyme-substrate reaction and persists only as long as the reaction is occurring. If either the substrate is used up or the enzyme looses activity then the reaction will cease and signal will be lost. However, in well-optimized assays using proper antibody dilutions and sufficient substrate, the reaction can produce stable output of light for 1 to 24 hours depending on the substrate, allowing consistent and sensitive detection that may be documented with X-ray film or digital imaging equipment. Of these detection methods, X-ray film is the most sensitive method in part because film is more densely coated with photo-reactive molecules than a digital imaging array has photon sensors. Also, film is placed nearly in direct contact with the blot, separated only by a thin transparent plastic sheet, allowing more light photons to collide with the film emulsion than with a digital sensor with is located behind one or more focusing lenses. For these reasons, film exposure times are shorter than those required for digital equipment. Unfortunately, some of the sensitivity will be lost when a digital scan must be made of film for publication purposes. Digital images circumvent this step by making a direct digital image of the blot, saving time in image processing, chemical waste and data manipulation.

The use of fluorophore-conjugated antibodies in a immunoassays requires fewer steps because there is no substrate development step in the assay. While the protocol is shorter, this method requires special equipment in order to detect and document the fluorescent signal due to the need for an excitation light source. Recent advances in digital imaging and develop of new fluorophore such as infrared, near-infrared and quantum dots has increased the sensitivity and popularity of using fluorescent probes for Western blotting and other immunoassays. Although the equipment and fluorescent-conjugated antibodies can be quite expensive, this method has the added advantage of multiplex compatibility (using more than one fluorophore in the same experiment). In addition, chemical waste is further reduced compared to other blotting procedures.

HYBRIDIZATION TECHNIQUES

Hybridization is the technique in which the nucleic acid or protein immobilized on the membrane is challenged with a probe or antibody known. Hybridization is widely used to confirm the presence or absence of the DNA/RNA/protein in the unknown sample. Hybridization depends on the function of the labelled base pair between the probe and the target sequence. After the blotting technique is completed, the membrane is placed in a solution of labelled (radioactive or non-radioactive) single stranded DNA or RNA solution. This DNA or RNA contains sequences complementary to DNA or RNA present on the membrane.

This labelled nucleic acid used to detect or locate DNA is called as probe. Conditions are chosen such that labelled DNA or RNA bind or hybridize with nucleic acid present on the membrane. If the sequence of nucleic acid in the probe is complementary to nucleotide sequence on the membrane then base pairing or hybridization will occur. The nucleic acid present on the membrane is single-stranded and is bound on the membrane by using negative charge of the phosphate and thymine molecules. Thus when it finds a complementary strand, it forms or develop hydrogen bonds or converts into hybrid DNA (i.e., double-stranded DNA in which the two strands come from different DNA molecules).

Conditions are chosen or maintained such that there is a maximum chance of specific hybridization and minimum of non-specific hybridization. After the hybridization, the membrane is washed to remove the unbounded probes, while bound probes remain attached. The regions of hybridization are detected by autoradiography method (if the probe is radioactively labelled) or by biotin streptavidin method (if the probe is labelled by a non-radioactive). The specificity with which a particular target sequence is detected by hybridization to a probe is called as stringency. At high stringency, hybridization occurs when the sequence on the membrane is completely complementary to the sequence of the probe. In practice, the hybridization is carried out at low stringency level, and the membrane is washed many times such that only perfect (>80%) matches are left over.

The principle of hybridization is more or less same for all, but to differentiate and make it conceptually undoubtful, various names are given. If the probes contain or are mode of DNA, they are called as DNA probes and if they are made of RNA they are called as RNA probes.If the membrane containing immobilized DNA is challenged with DNA probe, it is called as DNA hybridization. If the membrane containing immobilized DNA is challenged with RNA probe, then it is called as DNA: RNA hybridization. Similarly, if the membrane containing immobilized RNA is challenged with RNA probe then it is called as RNA hybridization and if challenged with DNA then it is called as RNA : DNA hybridization.

Applications of Blotting and Hybridization Techniques

1) Southern blotting technique is widely used to find specific nucleic acid sequence present in different animals including man. For example if we want to know whether there is a gene like insulin in sea anemone, then DNA of sea anemone is mobilized on membrane and blotted by using insulin probes against it.
2) Northern blotting technique is widely used to find gene expression and regulation of specific genes. For example if we find human insulin like gene in oyster, then by isolating and immobilizing RNA and blotting it with insulin probe we call tell whether the gene is expressing or not.
3) By using blotting technique we can identify infectious agents present in the sample.
4) We can identify inherited disease.
5) It can be applied to mapping restriction sites in single copy gene.

Disadvantages of Blotting and Hybridization Techniques

1) The process is a complex, cumbersome and time consuming one.

2) It requires electrophoretic separation.
3) Only one gene or RNA can be analysed at a time.
4) Gives information about presence of DNA, RNA or proteins but does not give information about regulation and gene interaction.

S1 MAPPING

The S1 nuclease is an endonuclease isolated from *Aspergillus oryzae* that digests single- but not double-stranded nucleic acid. In addition, it digests partially mismatched double-stranded molecules with such sensitivity that even a single base-pair mismatch can be cut and hence detected. In practice, a probe of end-labeled double-stranded DNA is denatured and hybridized to complementary RNA molecules. S1 is used to recognize and cut mismatches or unannealed regions and the products are analyzed on a denaturing polyacrylamide gel. A number of different uses of the S1 nuclease have been developed to analyze mRNA taking advantage of this property. Both qualitative and quantitative information can be obtained in the same experiment. The molecular weight of S1 nuclease is 29 kDa monomer.

Nuclease S1 will digest only ssDNA or ssRNA (Fig. 10.6). If a duplex of DNA and/or RNA strands has single stranded overhangs or unhybridized internal loops, these will be digested away. The remaining intact nucleic acid fragments represent regions of identity between two strands of the duplex. If one of the strands is labeled at one end, the length of labeled fragment remaining after hybridization and nuclease digestion reflects the point on the probe where the two sequences diverge. This is the basis for S-1 mapping of transcriptional start sites. A probe is chosen that is complementary to the RNA, and extends past the anticipated start site. The 5' end is ^{32}P labeled, and chosen to fall within the coding region of the mRNA, so that it will be protected from digestion. After hybridization, the 3' overhang of the probe is digested away, and the size of the remainder of the probe is accurately determined on a denaturing PAGE Gel. The distance between the known labeling site and the new end of the probe gives the transcriptional start site to within 1 base.

S1 mapping can also be used to find intron sites(Fig. 10.7). In this case, the probe is derived from genomic DNA, and again labeled so that the labeled 3' end falls within a coding portion of the gene. Any intron in this construct will not find a homologous region in the RNA, and will be cleaved by the S-1 nuclease. In this case, the size of the labeled remainder reflects the distance from the label site to the splice site.

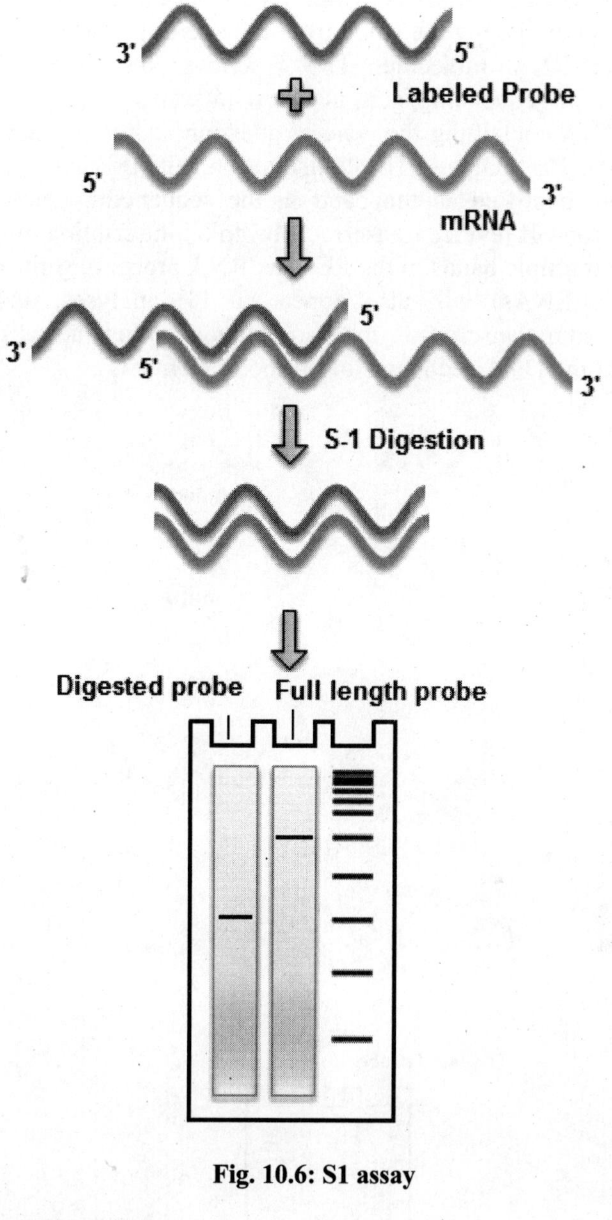

Fig. 10.6: S1 assay

PRIMER EXTENSION

The sites of transcription initiation or RNA processing are generally determined by S1 protection or primer extension (PE) analysis. Although S1 protection must be used for 3'-end analysis, 5'-end analysis is usually more easily & precisely done by primer extension analysis. In this method, a short antisense 5' end-labeled DNA primer (usually a synthetic oligonucleotide, but sometimes a small restriction fragment) is hybridized to RNA, usually total cellular RNA, then DNA is synthesized from this primer using reverse transcriptase. RT will copy the RNA from the site of primer annealing to the 5'-end of the RNA molecule. The reactions are then analyzed by electrophoresis in sequencing gels in lanes adjacent to Sanger sequencing reactions of DNA containing the gene of question, using the same primer as that used in the PE analysis. The transcription initiation site (usually) can then easily be identified as the band in the sequencing reaction directly parallel to the run-off reverse transcript. Multiple transcription initiation sites will appear as multiple bands in the PE lane. RNA processing sites (i.e. in the case of stable RNAs) will also appear in PE analyses, and must be distinguished from transcription initiation sites by other means (usually by examination of the DNA sequence for promoter elements).

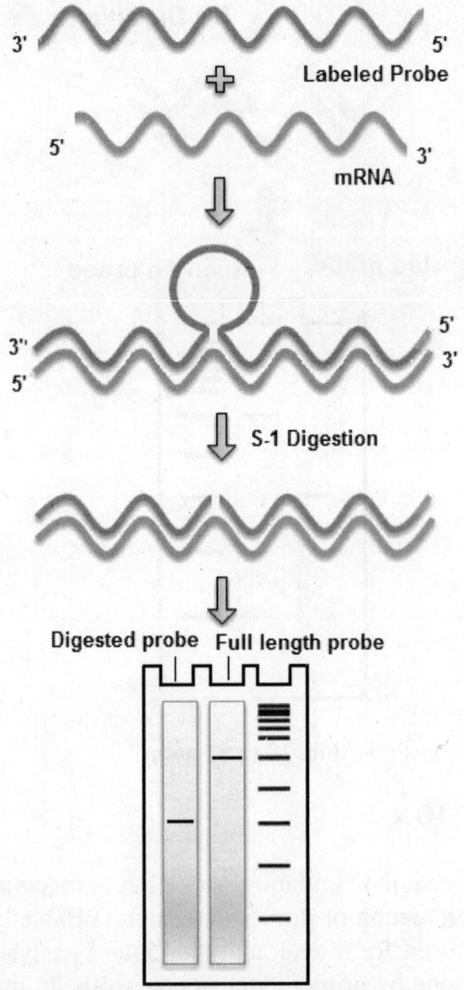

Fig. 10.7: Intron detection

Uses:

1) Mapping the 5' end of transcripts. This allows one to determine the start point of transcription (assuming the mRNA isn't further processed), which helps localize promoters or TATA boxes.
2) Quantifying the amount of transcript in an *in vitro* transcription system.
3) Determine the locations of breaks or modified bases in a mixed population of RNA or DNA samples. This is useful in applications like foot printing. Two different methods are used. In one, the modified nucleotide cannot be recognized by the polymerase or reverse transcriptase; in such cases, the chain ends at the site of modification. In the other, the modification is converted in a later step of the analysis to a strand break by chemical treatment. For instance, the sites of modifications by dimethyl sulfate (DMS) can be identified by treating DNA with DMS, exposing the sample to conditions that break the backbone at the site of modification, followed by primer extension.

RNASE PROTECTION ASSAY

Nuclease protection assays (NPAs), including both ribonuclease protection assays (RPAs) and S1 nuclease assays, are an extremely sensitive method for the detection, quantitation and mapping of specific RNAs in a complex mixture of total cellular RNA. The basis of NPAs is a solution hybridization of a single-stranded, discrete sized antisense probe(s) to an RNA sample. The small volume solution hybridization is far more efficient than more common membrane-based hybridization, and can accommodate up to 100 μg of total or poly(A) RNA. After hybridization, any remaining unhybridized probe and sample RNA are removed by digestion with a mixture of nucleases. Then, in a single step reaction, the nucleases are inactivated and the remaining probe: target hybrids are precipitated. These products are separated on a denaturing polyacrylamide gel and are visualized by autoradiography. If nonisotopic probes are used, samples are visualized by transferring the gel to a membrane and performing secondary detection.

NPAs are the method of choice for the simultaneous detection of several RNA species. During solution hybridization and subsequent analysis, individual probe/target interactions are completely independent of one another. Thus, several RNA targets and internal controls can be assayed simultaneously (up to twelve have been used in the same reaction), provided that the protected fragment of individual probes are of different lengths. NPAs are also commonly used to precisely map mRNA termini and intron/exon junctions.

RNA Quantitation

To quantitate mRNA levels using NPAs, the intensities of probe fragments protected by the sample RNA are compared to the intensities generated from either an endogenous internal control (relative quantitation) or known amounts of sense strand RNA (absolute quantitation).

Advantages of NPAs over Northern Blotting

- NPAs are more sensitive than traditional Northerns. They can be used to detect as little as 5 femtograms of target RNA or 4,000 to 50,000 copies/sample.
- NPAs are more tolerant of partially degraded RNA than Northerns. If samples are even slightly degraded, the quality of data from a Northern blot is severely compromised.
- NPAs are able to distinguish between transcripts of multi-gene families that may comigrate on Northerns.
- NPAs can be used to map mRNA termini and intron/exon junctions

Advantages of NPAs over RT-PCR Reactions

- RPAs are easy to use and don't require extensive optimization. Optimization reactions must be preformed prior to each RT-PCR reaction and the data obtained by RT-PCR analysis can be confusing and difficult to interpret.
- RPAs don't require expensive equipment purchases.

Limitations of Nuclease Protection Assays

The primary limitation of NPAs is the lack of information on transcript size. The portion of probe homologous to target RNA determines the size of the protected fragment. Another drawback to NPAs is the lack of probe flexibility. The most common type of NPA, the ribonuclease protection assay, requires the use of RNA probes. Oligonucleotides and other single-stranded DNA probes can only be used in assays containing S1 nuclease. A region of the single-stranded, antisense probe must typically be completely homologous to target RNA to prevent cleavage of the probe:target hybrid by nuclease. This means that partially related sequences (e.g., probe and target RNA from different species) usually cannot be used.

The RNase protection assay is a standard approach to determine mRNA levels of a gene of interest in different tissues, developmental stages, or times of the day. Splicing or promoter variants can be studied with specific probes.

It is widely used in chronobiology to study the temporal profile of expression of circadian genes and the effects of genetic manipulation on these oscillations.

RNase H

The enzyme **RNase H** (EC 3.1.26.4) is a ribonuclease that cleaves the 3'-O-P-bond of RNA in a DNA/RNA duplex to produce 3'-hydroxyl and 5'-phosphate terminated products. RNase H is a non-specific endonuclease and catalyzes the cleavage of RNA via an hydrolytic mechanism, aided by an enzyme-bound divalent metal ion. Members of the RNase H family can be found in nearly all organisms, from archaea and prokaryota to eukaryota. In DNA replication, RNase H is responsible for cutting out the RNA primer, allowing completion of the newly synthesized DNA. Retroviral RNase H, a part of the viral reverse transcriptase enzyme, is an important pharmaceutical target, as it is absolutely necessary for the proliferation of retroviruses, such as HIV. Inhibitors of this enzyme could therefore provide new drugs against diseases like AIDS. As of 2004, there are no RNase H inhibitors in clinical trials, though some approaches employing DNA aptamers are in the preclinical stage.

In a molecular biology laboratory, as RNase H specifically degrades the RNA in RNA:DNA hybrids and will not degrade DNA or unhybridized RNA, it is commonly used to destroy the RNA template after first-strand complementary DNA (cDNA) synthesis by reverse transcription, as well as procedures such as nuclease protection assays. RNase H can also be used to degrade specific RNA strands when the cDNA oligo is hybridized, such as the removal of the poly(A) tail from mRNA hybridized to oligo(dT), or the destruction of a chosen non-coding RNA inside or outside the living cell. To terminate the reaction, a chelator, such as EDTA, is often added to sequester the required metal ions in the reaction mixture.

REPORTER GENE ASSAYS

Reporter genes have become an invaluable tool in studies of gene expression. They are widely used in biomedical and pharmaceutical research and also in molecular biology and biochemistry. A gene consists of two functional parts: One is a DNA-sequence that gives the information about the protein that is produced (coding region). The other part is a specific DNA-sequence linked to the coding region; it regulates the transcription of the gene (promoter). The promoter is either activating or suppressing the expression of the gene. The purpose of the reporter gene assay is to measure the regulatory potential of an unknown DNA-sequence. This can be done by linking a promoter sequence to an easily detectable reporter gene such as that encoding for the firefly luciferase.

Common reporter genes are β-galactosidase, β-glucuronidase and luciferase. Various detection methods are used to measure expressed reporter gene protein. These include luminescence, absorbance and fluorescence.

In molecular biology, a reporter gene (often simply reporter) is a gene that researchers attach to a regulatory sequence of another gene of interest in cell culture, animals or plants. Certain genes are chosen as reporters because the characteristics they confer on organisms expressing them are easily identified and measured, or because they are selectable markers. Reporter genes are generally used to determine whether the gene of interest has been taken up by or expressed in the cell or organism population.

To introduce a reporter gene into an organism, scientists place the reporter gene and the gene of interest in the same DNA construct to be inserted into the cell or organism. For bacteria or eukaryotic cells in culture, this is usually in the form of a circular DNA molecule called a plasmid(Fig. 10.8). It is important to use a reporter gene that is not natively.

expressed in the cell or organism under study, since the expression of the reporter is being used as a marker for successful uptake of the gene of interest. The different Reporter genes used in the recombinant DNA technology are isolated from different sources and used for the different constructs for plant, animal and microbial cultures. The different reporter genes used are as follows:

Bacterial β-galactosidase

Bacterial **β-galactosidase** catalyzes the hydrolysis of β-galactosides. This enzyme is encoded by *lacZ* of *Escherichia coli* and can be used in prokaryotic as well as in eukaryotic cells. The enzyme has a high turnover rate and generates strong signals by using fluorescence, electrochemical or chemiluminescence substrates.

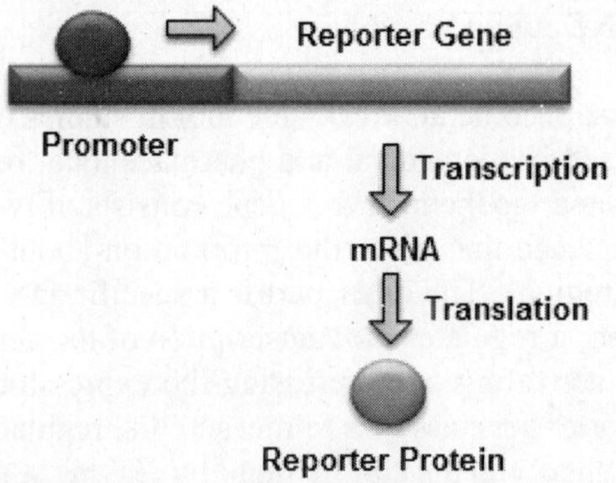

Fig. 10.8: Reporter gene assay

The advantages of a luciferase assay are the high sensitivity, the absence of luciferase activity inside most of the cell types, the wide dynamic range, rapidity and low costs. The firefly luciferase catalyzes the bioluminescent oxidation of the luciferin in the presence of ATP, Magnesium and Oxygen.

The most versatile and common reporter gene is the luciferase of the North American firefly *Photinus pyralis.* The protein requires no posttranslational modification for enzyme activity; it is not even toxic in high concentration (in vivo) and can be used in pro- and eukaryotic cells.

GFP, green fluorescent protein

GFP, **green fluorescent protein**, isolated from the jellyfish *Aequorea victoria*, fluoresces in this animal due to an energy transfer from the Ca2+-activated photoprotein aequorin. The highly stable GFP protein has a high quantum yield (0.88), and can be expressed in both prokaryotic and eukaryotic systems with no need of a substrate or cofactor. GFP does not have any enzymatic amplification, thus reducing its potential detection sensitivity. It has been shown that 105 to 106 GFP molecules are needed to allow detection over background fluorescence in a single cell.

β-glucuronidase

β-glucuronidase (GUS) is mainly used for plant cells and there are some reports of its application in mammalian systems. Different substrates are known, forming colored, fluorescent, or luminescent cleavage products; the sensitivity of the luminescent assay is the highest.

Alkaline phosphatase

Alkaline phosphatase (AP) and secreted alkaline phosphatase (SEAP) are orthophosphoric monoester phosphohydrolases with an alkaline pH optimum. These stable enzymes are characterized by a high turnover rate. The classical determination is colorimetric, but extremely sensitive fluorescent and luminometric detection systems were reported; the latter allow a sensitivity similar or better compared to that of luciferase.

Neomycin Phosphotransferase Gene (nptll)

This gene is used both as a selectable and a scoreable marker in experiments involving transfer of genes leading to the production of transgenic plants. It imparts kanamycin resistance, so that the transformed tissue can be selected on kanamycin. An assay for NPTII enzyme is also used to detect its presence in transformed tissue or transgenic plants. The gene for NPTII enzyme is often used with *nos* promoter, which drives its synthesis. In some cases, nptll gene had adverse effect on theexpression of the desirable gene introduced, so that alternative approaches for improving itsexpression had to be used.

Chloramphenicol Acetyl Transferase Gene (cat)

This gene is primarily used as a reporter gene or a scoreale marker and not as a selectable marker. The gene first isolated from *E. coli* codes for an enzyme chloramphenicol acetyl transferase (CAT), which is absent in mammals and higher plants, so that whenever the gene *cat* is transferred with a gene construct, its presence can be detected by enzyme assay. Rarely screening for this enzyme may also be used for selection of transformed regenerants, although no selection pressure can be applied. The enzyme uses acetyl CoA + chloramphenicol ^{32}P)as sub states and helps in the transfer of acetyl group to chloramphenicol. The presence of acetyl chloramphenicol is detected through autoradiography: The gene *cat* has also been used for identification of a number of regulatory sequences.

Phosphomannose Isomerase Gene (pmi)

Although the genes for antibiotic and herbicide resistance as selectable markers for plant transformation have been shown to be completely safe and seem to pose no risk to the farmers or consumers, public concerns have been voiced against these markers, particularly when used in food crops. Therefore, efforts have been made to use systems, which will remove the selectable marker, after the selection of transformed cells is over. However, to address public concerns, efforts are also being made to phase out the use of antibiotic resistance genes as markers. As a step in this direction, during 2000-2001, a company named Novartis developed and used a selectable marker system that makes use of phosphomannose isomerase gene (pmi)derived from *E. coli.* The product of this gene, the enzyme phosphomannose isomerase (PMI) converts mannopse-6-phosphate (obtained from mannose due to action of hexokinase) to fructose-6- phosphate, which can be utilized by transformed cultured cells. The method for the use of this marker system in crop plants involves cells or tissues using any method, followed by placement of these cells in a culture medium, to which mannose is added as a substrate either as a sole carbon source or in combination with sucrose. Mannose has no adverse effect on plant cells, but selection occurs due to its phosphorylation to monnose-6- phosphate by hexokinase. In tissues lacking PMI, mannose-6-phosphate accumulates and the cells stop growing, thus offering selection of transformed cells, which can metabolise mannose-6-phospote due to the presence of gene pmi.

The expression of reporter genes is easily detected either through highly sensitive enzyme assays (scorable reporter genes) or through expression of resistance to a toxin (selectable reporter genes). Some commonly used easily detectable enzyme producing genes are, nos (nopaline synthase, from *Agrobacterium*), lux (luciferase from bacteria or firefly), cat (chloramphenicol acetyltransferase from bacteria), and gus (β-glucuronidase from bacteria), etc. Activities of the enzymes produced by scorable reporter genes are determined either in situ, i.e., in the transformed tissues, or by in vitro assays using plant tissue extracts. In addition, immunological methods may also be used to detect the protein products of marker genes either in situ, in plant extracts or by western blotting.

The essential features of an ideal reporter gene are:

1) Lack of endogenous activity in plant cells of the concerned enzyme,
2) An efficient and easy detection, and
3) A relatively rapid degradation of the enzyme.

Each reporter gene has some advantages and some disadvantages and none of them are ideally suited for all plant species. For example, there is little or no endogenous luciferase activity in plant cells, the enzyme serves as a visual marker, the assay is quite sensitive, but the enzyme is highly stable. Reporter genes, more particularly, scorable reporter genes have been extensively used to assay the function of promoters and other regulatory sequences, and also to demonstrate the transformation of different plants, animal or microbial cells. Genetic reporters are used as indicators to study gene expression and cellular events coupled to gene expression. They are widely used in pharmaceutical and biomedical research and also in molecular biology and biochemistry. Typically, a reporter gene is cloned with a DNA sequence of interest into an expression vector that is then transferred into cells. Following transfer, the cells are assayed for the presence of the reporter by directly measuring the reporter protein itself or the enzymatic activity of the reporter protein. A good reporter gene can be identified easily and measured quantitatively when it is expressed (in the organism or cells of interest).

11

Production of Recombinant Proteins

Molecular biotechnology is an exciting revolutionary scientific discipline that is based on the ability of a researcher to transfer specific units of genetic information from one organism to another. The objective of recombinant DNA technology is often to produce a useful product or a commercial process. In early 1970, traditional biotechnology was not well known as a scientific discipline. Research in this area was carried out in chemical engineering departments. The term "biotechnology" was created in 1919 by a Hungarian engineer, Karl Ereky. According to Ereky, "biotechnology involves all works carried out with the aid of living things." More formally, biotechnology may be defined as "the application of scientific and engineering principles to the processing of material by biological agents to provide goods and services."

Genetic engineering provided a means to create, rather than merely isolate highly productive strains. Microorganisms and eukaryotic cells could be used as "biological factories" for the production of insulin, interferon, growth hormone, viral antigens and a variety of other proteins. Recombinant DNA technology could also be used to facilitate the biological production oflarge amounts of useful lower molecular weight compounds and macromolecules that occur naturally in minute quantities. This, technology facilitated the development of radically new medical therapies and diagnostic systems. Thus, the union of recombinant DNA technology with biotechnology created a vibrant, highly competitive field of study that has been called "Molecular Biotechnology."Molecular biotechnology ought to contribute unprecedented benefits to humanity.

1) It should provide opportunities to accurately diagnose and prevent or cure a wide range of infectious and genetic diseases.

2) Significant increase in crop yield may be obtained by generating disease-, pathogen- or herbicide-resistant varieties.

3) Microorganisms that will produce chemicals, antibiotics, polymers, amino acids, enzymes, etc. can be developed.

4) Livestock and other animals that have enhanced genetically determined attributes can be developed. Molecular biotechnology with much fuss and fanfare has become a comprehensive scientific venture, both commercially and academically, in a remarkably short time.

5) A number of new scientific and business publications are devoted to molecular biotechnology. Both graduate and undergraduate programmes and courses have been created at many universities throughout the world.

RECOMBINANT PROTEINS

Proteins produced by genes transferred into selected host cells by genetic engineering are called recombinant proteins since they are based on recombinant DNA technology (Fig. 11.1). Recombinant proteins form an important component of biopharmaceuticals, i.e., biotechnology products having pharmaceutical applications. A large number (nearly 2 dozen) of recombinant proteins are being produced in mammalian cell cultures, some of which, viz., human growth hormone (hGH), tissue plasminogen activator (tPA), erythropoietin and blood clotting factor VIII, are already in therapeutic use. The other products are in advanced stages of development. The host cells used for large scale production of the various recombinant proteins are Chinese hamster ovary cell line (CHO), baby hamster kidney line BHK, mouse mammary tumour line C127, mouse myeloma cell lines and mammalian cell lines. The use of animal cell lines has been made possible by the following developments in cell culture technology:

1) Development of culture systems permitting large scale culture of animal cells at high densities.

2) Development of media, which minimize or even obviate the use of serum (serum interferes with downstream processing, i.e., separation and purification of the desired product).

Genetic engineering has expanded the industrial applications of microorganisms including production of human proteins. By using recombinant DNA technology, human DNA sequences that code for various proteins have been incorporated into the genomes of bacteria. By growing these recombinant bacteria in fermentors, human proteins could be produced commercially.

ADVANTAGES OF RECOMBINANT PROTEINS

Production of recombinant proteins of pharmaceutical value in microorganisms, cultured animal cells or plants offers several important advantages over their conventional routes of production.

1) Production costs are reduced, in some cases drastically.
2) There is no risk of contamination by AIDS or any other virus.\
3) They can be produced in far greater quantities than it is possible for conventional products.
4) In some cases, proteins of human origin have become available, e.g., insulin.

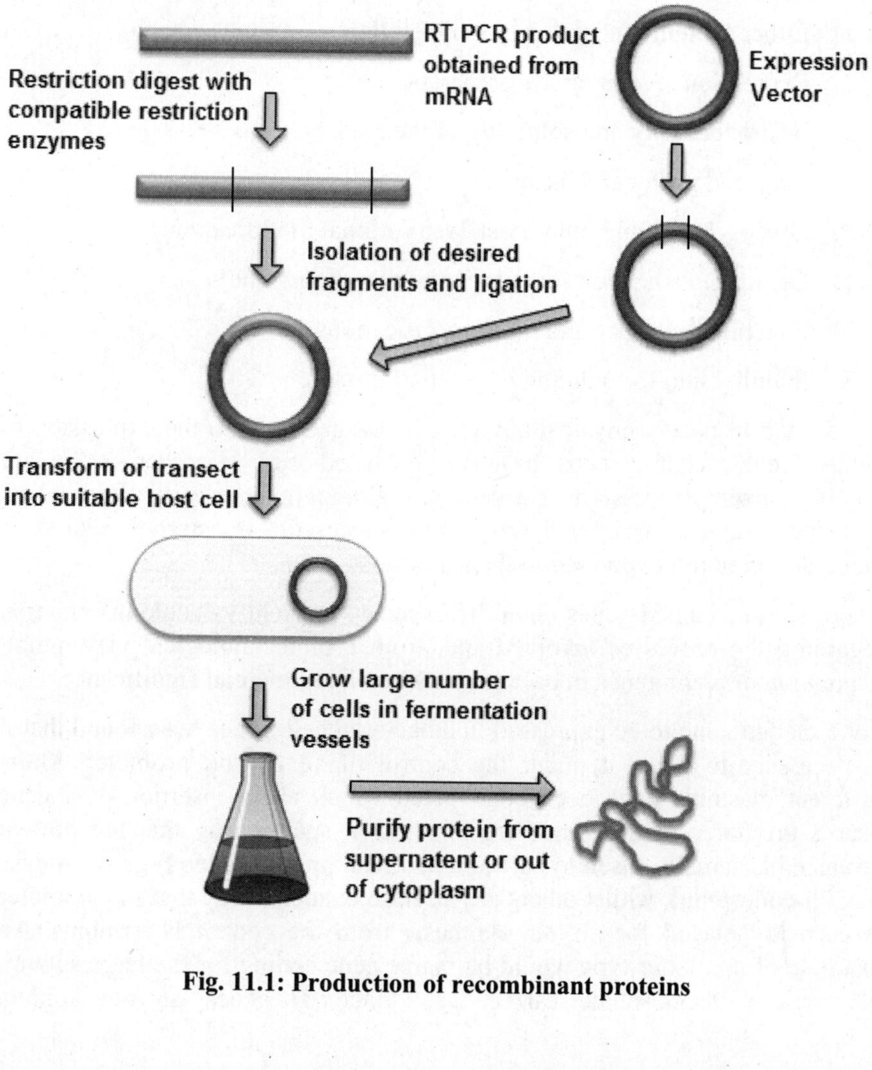

Fig. 11.1: Production of recombinant proteins

PROTEIN EXPRESSION

Protein expression and purification have neither a fixed approach nor a protocol for all. Every project of expressing and purifying a protein is unique and requires specific analyses, from bioinformatics analysis and engineering design of protein sequence, through selection of expression vector and host system, to meeting various requirements or purposes of the final product (expressed and purified protein). Compared with recombinant DNA technologies, protein expression and purification technologies are much more complicated and sophisticated. This is why there are so many methods and protocols developed for successfully carrying out a recombinant protein project. It depends on the various factors related to the protein structure and function. Following are some major points to be considered while expressing a protein from one species to another:

1) Either protein is secreted or intracellular.
2) Expression level in crude preparation.
3) Hydrophobicity and solubility of the protein.
4) Required purity and quantity.
5) Glycosylation and other Post Translational modification..
6) Cellular localization (membrane or non-membrane).
7) Functional activity and other specifications.
8) Stability and formulation of purified protein.

In addition to purely physical analysis, investigations into the expression of cloned genes in host cells have been carried out. Many processes are involved in the expression of a gene and although some early experiments with hosts such as yeast and fungi were successful, transferred eukaryotic genes are often not expressed in their new host.

Expression of cloned genes could be expected to yield valuable information regarding the processes involved and, from a biotechnological view point, expression of plant genes in bacteria may be of commercial significance.

For a cloned gene to be expressed in a bacterial cell, it has been found that it is necessary to place it under the control of an *E. coli* promoter. Many different plasmids have been constructed which allow insertion of a gene near a promoter. Some genes are inserted in such a way that the protein produced is actually fused to part of a bacterial protein, often β-galactosidase (e.g., β-endorphin), whilst others are inserted in such a way that the promoter is correctly placed for protein synthesis from the correct N-terminus. An example of this latter type would be maize gene coding for the large subunit of ribulose bisphosphate carboxylase. Bacterial genes do not contain intervening sequences nor the machinery for removing them from the primary transcript.

Eukaryotic genes containing introns will not give a functional mRNA molecule in bacteria even if transcribed efficiently. It is therefore, necessary to use cDNA clones of this type of gene for expression studies (e.g., sweet protein, thaumatin). A functional mRNA also depends on the coding sequence being in the correct reading frame if it has been fused to a bacterial coding sequence. Three vectors were constructed by Charnay et al. (1978) which allow cloning into the lac-Z gene of *E. coli* in all three possible reading frames relative to the initiation codon. Similar vectors have been constructed by a number of other workers. If placed in the correct reading frame relative to the initiation codon and with a ribosome binding site present, the mRNA should be translated into a primary translation product. The problems do not end there, however; many proteins are modified in some way either by removal of the polypeptide or by addition of various groups. Signal sequences, which allow passage through membranes, need to be cloven and other modifications include glycosylation, adenylation and phosphorylation. While bacteria may be able to cleave signal sequences, in some cases glycosylation certainly does not occur. In the case of the sweet protein, thaumatin, neither the N- nor C-terminal extensions of the primary translation product were removed in E. coli. Likewise, the glutamine synthetase gene from Anabaena functions in E. coli but no adenylation of the enzyme occurs. Lack of these modifications may have certain consequences. If a functional protein is sought for some commercial application, lack of these modifications could be a serious problem. From an investigative point of view, the primary products may be very unstable. If rapid degradation occurs, detection of the products would be difficult. Expression of a cloned gene requires correct functioning of a complete series of events. A eukaryotic host, such as yeast, may have considerable advantages for some application as modification of primary translation products is known to occur in these organisms.

EXPRESSION IN EUKARYOTES

Cloning in Eukaryotes

Since chromosomes found in the nucleus of eukaryotes are separated from the rest of cell through nuclear membrane, and since many of the genes are split genes with exons and introns, genetic engineering with eukaryotes requires new methods and tools.

When eukaryotic genes are cloned in prokaryotes, the split genes can not be correctly expressed, because prokaryotes do not have the equipment necessary for splicing out the RNA transcribed from the introns of a gene. In view of this, eukaryotic cells may sometimes be needed for cloning and particularly for expression of cloned eukaryotic genes. Among eukaryotes, DNA cloning has been done in yeast, mouse and to some extent even in some higher plant species. In yeast, a plasmid called 2p DNA (63 bp) is found, which is an appropriate cloning vehicle. An efficient transformation method is also available, which involves protoplast production followed by PEG directed introduction of DNA into protoplasts.

In animal cells like mouse cells, special animal viruses were used as cloning vehicles. Simian virus 40 (SV4O) is one such virus, in which globin gene could be integrated. This gene integrated in SV 40 could be transcribed and translated in mouse kidney cells. A variety of human tell lines in culture is also used for cloning of foreign DNA, using some animal viruses as vectors.

Expression strategies for heterologous Genes

Heterologous gene expression the synthesis of foreign proteins in a host organism following transformation of that organism by a vector carrying genes from a different organism. However, this can pose a problem in bacterial hosts, which can degrade the heterologous proteins by proteases. Some heterologous proteins are deposited into insoluble inclusion bodies while others fail to fold properly. Furthermore, bacteria are incapable of adding sugar residues to proteins that require glycosylation. This problem can be solved by the vector and host engineering before the cloning and expression of heterologous genes.

Expression of Vectors for High Level of Expression of Cloned Genes

In a number of cloning vectors the main utility has been to clone a gene or a DNA sequence. In other words, they are used for obtaining millions of copies of the cloned DNA segment, for further use in either the genetic engineering experiments or for further basic studies. With this objective in mind, the cloned genes in these vectors need not express themselves either at the transcription level (production of mRNA) or at the translation level (production of protein). But, when the cloned gene is used for transformation to generate transgenic plants or animals or for production of microbes to be used in industry, then the cloned gene must be expressed. Sometimes, a high level of expression of gene is desirable, if the product of cloned gene is to be recovered as a commercial product. This objective can be achieved through the use of promoters and expression cassettes (an expression cassette is a combination of DNA sequences, which allow faithful expression of the cloned gene).

Promoters

Promoters from non plant sources express poorly in plants (except promoters of T -DNA) and promoters of non-animal sources express poorly in animals.

Therefore, to obtain correct expression, genes transferred to plants should be linked to plant specific promoters and those transferred to animals should be linked to animal specific promoters. These promoters can also be interchanged so as to confer on the transferred gene, a specific pattern of expression (tissue specific or stimulus specific or developmental stage specific expression). To illustrate this, some examples of promoters used in plants and insects will be discussed.

Nopaline Synthase (nos) Promoter From T-DNA

This promoter is 200 bp long and contains several DNA sequence motifs which direct the expression of linked gene. The upstream element from - 97 to - 130, when duplicated, increases the expression three fold, suggesting that this may be an enhancer. The *nos* promoter is most active in basal regions of plants, its activity slowing down in vegetative parts at the onset of flowering, even though in the flower its activity is increased.

Dual Promoter of Mannopine Synthase (mas) Genes 1 and 2

These promoters are present in a T-DNA fragment 467 bp long. The expression seems to be developmentally regulated and the linked genes seem to be most active in the basal region of the plant. Expression is also induced on wounding. Since they are closely linked, if a marker gene is linked to one of these two promoters and a foreign gene is linked to the other promoter, selection for high expression of marker gene will lead to high level of expression of the foreign gene.

35S RNA Promoter od CaMV

This is the most extensively used promoter in a wide variety of plants. The promoter is 343 bp long and contains a strong transcriptional enhancer, which on duplication results in 10 fold increase in expression, even at a distance of 2 kbp. High expression is mainly observed in leaf tissue.

Polyhedrin Promoter From Baculovirus

Baculovirus can synthesize polyhedrin protein in infected cells (baculovirus is an insect virus), so that polyhedrin makes upto 50% of the total protein in the cell, giving upto 1g of protein per liter of insect cell culture. In view of this, the promoter has been used to construct expression vectors to allow high level of expression of any gene under the influence of this promoter. This is being used for developing biopesticides or even for production of specific chemicals in industry.

Expression Cassettes

The literal meaning of cassette is a device containing film such as magnetic tape for insertion into equipment like tape recorder, VCR, etc. Therefore by expression cassette, we mean gene constructs, which allow the insertion of foreign genes, either as transcriptional or translational fusions, behind specific promoters. pRT plasmids are typical in this connection and have been derived from pUCI8/19. They consist of a series of vectors (e.g. pRT100, pRT101, pRT102, pRT103, pRT104), which differ in polylinker sequences, each flanked by 35S promoter of CaMV on one end and a sequence for poly A addition on the other end. A variety of marker genes (cat, nptII, gus, etc.) have been inserted into these cassettes and their expression studied, both in protoplasts (transient expression) and in stable transgenic tissues.

Baculovirus and Expression Vector System for Insect Cells

Development of baculovirus expression vector system is based on its life cycle which involves the following steps:

(i) In early phase (12 hours), virus particles are budded from infected cells and spread the virus.

(ii) In the late phase (2-5 days), occluded virus units accumulate in the nucleus and get embedded in protein polyhedra.

(iii) Insect is decomposed releasing polyhedra.

(iv) Polyhedra are ingested by another insect, where polyhedra are dissolved releasing virus for multiplication

The use of baculovirus as a cloning and expression vector involves following steps:

(i) Desired gene is cloned in a transfer vector under the control of polyhedrin promoter.

(ii) Chimeric transfer vector is contransfected with wild type virus into the host cell.

(iii) Polyhedrin gene of wild type is replaced by cloned gene through recombination.

(iv) Recombinant virus is plaque purified, since they will have no polyhedra occlusion bodies.

(v) The selected recombinant virus is multiplied and used for expression of heterologous protein.

Codon Optimization

The genetic code (at first approximation) uses 64 codons to encode 21 different signals; hence there are some choices as to which codon to use. Amino acids and stop can have 1,2,3,4 or 6 codons in the standard scheme of things. But, those codons are rarely used with equal frequency. Leucine, for example, has 6 codons and some are rarely used and others often. Which codons are preferred and disfavored, and the degree to which this is true, depends on the organism. In the extreme, a codon can actually go so out of favor it goes extinct & can no longer be used, and sometimes it is later reassigned to something else; hence some of the more tidy codes in certain organisms.

A further observation is that the more favored codons correspond to more abundant tRNAs and less favored ones to less abundant tRNAs. Furthermore, highly expressed genes are often rich in favored codons and lowly expressed ones much more likely to use rare ones. To complete the picture, in organisms such as E.coli there are genes which don't seem to follow the usual pattern -- and these are often associated with mobile elements and phage or have other suggestions that they may be recent acquisitions from another species. A practical application of this is to codon optimize genes. If a gene built to express a protein in a foreign host, then it would seem apropos to adjust the codon usage to the local dialect, which usually still leaves plenty of room to accommodate other wishes (such as avoiding the recognition sites for specific restriction enzymes). There are at least four major schemes for doing this, with different gene synthesis vendors preferring one or the other.

CAI Maximization

CAI is a measure of usage of preferred codons; this strategy tries to maximize the statistic by using the most preferred codons. Therefore if these are the most preferred codons, and highly expressed genes are rich in them, why not do the same?

Codon sampling

This strategy (which is what Codon Devices offered) samples from a set of codons with probabilities proportional to their usage in the organism, after first zeroing out the very rare codons and renormalizing the table. So it avoids the rare ones, but don't hammer the better ones either; balance is always good

Dicodon optimization

In addition to codons showing preferences, there's also a pattern by which adjacent codons pair slightly non-randomly. One particular example; very rare codons are very unlikely to be followed by another very rare codon.

Codon frequency matching

Roughly, this means look at the native mRNA and its uses of codons and ape this in the target species; a codon which is rare in the native should be replaced with one rare in the target. As some rare codons may just help fold things properly.

In vitro Transcription

The ability to synthesize RNA in the laboratory is critical to many techniques. Radiolabeled and nonisotopically labeled RNA probes, generated in small scale transcription reactions, can be used in blot hybridizations and nuclease protection assays. Such probes are much more sensitive than random-primed DNA probes. Small scale reactions may also be used to synthesize RNA transcripts containing modified nucleotides for various biochemical and molecular biology studies. Large scale transcription reactions, generating up to 200 μg of RNA per reaction can be used for RNA amplification, expression studies (microinjection, infection with viral transcripts, in vitro translation), structural analysis (protein-RNA binding), and mechanistic studies (ribozyme analyses).

Requirements for Transcription

In vitro transcription requires a purified linear DNA template containing a promoter, ribonucleotide triphosphates, a buffer system that includes DTT and magnesium ions, and an appropriate phage RNA polymerase. The exact conditions used in the transcription reaction depend on the amount of RNA needed for a specific application.

RNA Phage Polymerases

The common RNA polymerases used in *in vitro* transcription reactions are SP6, T7 and T3 polymerases, named for the Bacteriophages from which they were cloned. The genes for these proteins have been over expressed in *E. coli* and the polymerases have been rigorously purified. RNA polymerases are DNA template-dependent with distinct and very specific promoter sequence requirements. After the RNA polymerase binds to its double-stranded DNA promoter, the polymerase separates the two DNA strands and uses the 3' - 5' strand as template for the synthesis of a complementary 5' - 3' RNA strand. Depending on the orientation of cDNA sequence relative to the promoter, the template may be designed to produce sense strand or antisense strand RNA. Although SP6 polymerase is often perceived to be less efficient than T7 or T3 polymerases, under optimal reaction conditions but it is found that all three RNA polymerases synthesize RNA with roughly equal efficiency; Differences in transcription can often be explained by differential sensitivity to salt. SP6 polymerase is the most sensitive to salt contamination often carried over with the DNA template.

Template Options: Plasmids, PCR Products, Oligonucleotides and cDNA

The DNA template must contain a double-stranded promoter region where the phage polymerase binds and initiates RNA synthesis. Transcription templates include plasmid constructs engineered by cloning, cDNA templates generated by first- and second-strand synthesis from an RNA precursor (e.g.,

RNA amplification), and linear templates generated by PCR or by annealing chemically synthesized oligonucleotides.

Plasmids

Many common plasmid cloning vectors include phage polymerase promoters. They often contain two distinct promoters, one on each side of the multiple cloning site, allowing transcription of either strand of an inserted sequence. Plasmid vectors used as transcription templates should be linearized by restriction enzyme digestion. Because transcription proceeds to the end of the DNA template, linearization ensures that RNA transcripts of a defined length and sequence are generated. The restriction site need not be unique, and providing the promoter remains adjacent to the transcription template, the vector itself may be digested multiple times. It is also unnecessary to purify the promoter-insert sequence away from other fragments prior to transcription because only the fragment containing promoter sequence will serve as template. Restriction enzyme digestion should be followed by purification since contaminants in the digestion reaction may inhibit transcription.

PCR Products

PCR products can also function as templates for transcription. A promoter can be added to the PCR product by including the promoter sequence at the 5' end of either the forward or reverse PCR primer. These bases become double-stranded promoter sequence during the PCR reaction.

Oligonucleotides

Two oligonucleotides can also be used to create short transcription templates. Two complementary oligonucleotides containing a phage promoter sequence are simply annealed to make a double-stranded DNA template. Only part of the DNA template - the -17 to +1 bases of the RNA polymerase promoter - needs to be double-stranded. It may be more economical, therefore, to synthesize one short and one long oligonucleotide, generating an asymmetric hybrid.

cDNA

A more recent use of in vitro transcription is in aRNA amplification reactions. For these reactions, transcription templates are generated from RNA by using an oligo (dT)-T7 promoter primer during reverse transcription. The cDNA is converted to a double-stranded transcription template by a second-strand synthesis reaction.

Sense or Antisense

When designing a transcription template, it must be decided whether sense or antisense transcripts are needed. If the RNA is to be used as a probe for hybridization to messenger RNA (e.g. Northern blots, in situ hybridizations, and nuclease protection assays), complementary antisense transcripts are required. In contrast, sense strand transcripts are used when performing expression, structural or functional studies or when constructing a standard curve for RNA quantitation using an artificial sense strand RNA. The +1 G of the RNA polymerase promoter sequence in the DNA template is the first base incorporated into the transcription product. To make sense RNA, the 5' end of the coding strand must be adjacent to or just downstream of, the +1 G of the promoter. For antisense RNA to bc transcribed the 5' end of the noncoding strand must be adjacent to the +1 G. If the insert is in a vector, the vector should be linearized dcwnstream from the promoter and the inserted sequence to be transcribed.

Conventional or Large Scale Synthesis

In vitro transcription reactions can be divided into two types: conventional and large scale. Conventional reactions are typically used for synthesizing radiolabelled RNA probes or for incorporating modified nucleotides into transcripts. Large scale reactions, which generate >100 μg RNA per reaction, are useful for structural and expression studies, as well as for RNA amplification.

In vitro translation

The in vitro synthesis of proteins in cell-free extracts is an important tool for molecular biologists and has a variety of applications, including the rapid identification of gene products (e.g., proteomics), localization of mutations through synthesis of truncated gene products, protein folding studies, and incorporation of modified or unnatural amino acids for functional studies. The use of in vitro translation systems can have advantages over in vivo gene expression when the over-expressed product is toxic to the host cell, when the product is insoluble or forms inclusion bodies, or when the protein undergoes rapid proteolytic degradation by intracellular proteases. In principle, it should be possible to prepare a cell-free extract for in vitro translation of mRNAs from any type of cells. In practice, only a few cell-free systems have been developed for in vitro protein synthesis. In general, these systems are derived from cells engaged in a high rate of protein synthesis.

Cell-Free Expression Systems

The most frequently used cell-free translation systems consist of extracts from rabbit reticulocytes, wheat germ and *Escherichia coli*. All are prepared as crude extracts containing all the macromolecular components (70S or 80S ribosomes, tRNAs, aminoacyl-tRNA synthetases, initiation, elongation and termination factors, etc.) required for translation of exogenous RNA. To ensure efficient translation, each extract must be supplemented with amino acids, energy sources (ATP, GTP), energy regenerating systems (creatine phosphate and creatine phosphokinase for eukaryotic systems, and phosphoenol pyruvate and pyruvate kinase for the *E. coli* lysate), and other co-factors (Mg^{2+}, K^{+}, etc.). There are two approaches to in vitro protein synthesis based on the starting genetic material: RNA or DNA. Standard translation systems, such as reticulocyte lysates and wheat germ extracts, use RNA as a template; whereas "coupled" and "linked" systems start with DNA templates, which are transcribed into RNA then translated. Each of these systems is discussed below.

Rabbit Reticulocyte Lysate

Rabbit reticulocyte lysate is a highly efficient in vitro eukaryotic protein synthesis system used for translation of exogenous RNAs (either natural or generated in vitro). In vivo, reticulocytes are highly specialized cells primarily responsible for the synthesis of hemoglobin, which represents more than 90% of the protein made in the reticulocyte. These immature red cells have already lost their nuclei, but contain adequate mRNA, as well as complete translation machinery, for extensive globin synthesis. The endogenous globin mRNA can be eliminated by incubation with Ca^{2+}-dependent micrococcal nuclease, which is later inactivated by chelation of the Ca^{2+} by EGTA. Nuclease-treated reticulocyte lysate is the most widely used RNA-dependent cell-free system because of its low background and its efficient utilization of exogenous RNAs even at low concentrations. Exogenous proteins are synthesized at a rate close to that observed in intact reticulocyte cells. Untreated reticulocyte lysate translates endogenous globin mRNA, exogenous RNAs, or both. This type of lysate is typically used for studying the translation machinery, e.g. studying the effects of inhibitors on globin translation. Both the untreated and treated rabbit reticulocyte lysates have low nuclease activity and are capable of synthesizing a large amount of full-length product. Both lysates are appropriate for the synthesis of larger proteins from either capped or uncapped RNAs (eukaryotic or viral).

Wheat Germ Extract

Wheat germ extract is a convenient alternative to the rabbit reticulocyte lysate cell-free system. This extract has low background incorporation due to its low level of endogenous mRNA. Wheat germ lysate efficiently translates exogenous RNA from a variety of different organisms, from viruses and yeast to higher plants and mammals. The wheat germ extract is recommended for translation of RNA containing small fragments of double-stranded RNA or oxidized thiols, which are inhibitory to the rabbit reticulocyte lysate. Both retic and wheat germ extracts translate RNA isolated from cells and tissue or those generated by in vitro transcription. When using RNA synthesized in vitro, the presence of a 5' cap structure may enhance translational activity. Typically, translation by wheat germ extracts is more cap-dependent than translation by retic extracts. If capping of the RNA is impossible and the protein yield from an uncapped mRNA is low, the coding sequence can be subcloned into a prokaryotic vector and expressed directly from a DNA template in an *E.coli* cell-free system.

E. coli Cell-Free System

E. coli cell-free systems consist of a crude extract that is rich in endogenous mRNA. The extract is incubated during preparation so that this endogenous mRNA is translated and subsequently degraded. Because the levels of endogenous mRNA in the prepared lysate is low, the exogenous product is easily identified. In comparison to eukaryotic systems, the *E.coli* extract has a relatively simple translational apparatus with less complicated control at the initiation level, allowing this system to be very efficient in protein synthesis. Bacterial extracts are often unsuitable for translation of RNA, because exogenous RNA is rapidly degraded by endogenous nucleases. There are some viral mRNAs (TMV, STNV, and MS2) that translate efficiently, because they are somewhat resistant to nuclease activity and contain stable secondary structure. However, *E.coli* extracts are ideal for coupled transcription: translation from DNA templates.

"Linked" And "Coupled" Transcription :Translation Systems

In standard translation reactions, purified RNA is used as a template for translation. "Linked" and "coupled" systems, on the other hand, use DNA as a template. RNA is transcribed from the DNA and subsequently translated without any purification. Such systems typically combine a prokaryotic phage RNA polymerase and promoter (T7, T3, or SP6) with eukaryotic or prokaryotic extracts to synthesize proteins from exogenous DNA templates. DNA templates for transcription: translation reactions may be cloned into plasmid vectors or generated by PCR.

Linked Transcription: Translation

The "linked" system is a two-step reaction, based on transcription with a bacteriophage polymerase followed by translation in the rabbit reticulocyte lysate or wheat germ lysate. Because the transcription and translation reactions are separate, each can be optimized to ensure that both are functioning at their full potential. Conversely, much commercially available eukaryotic coupled transcription: translation systems have compromised one or both reactions so that they can occur in a single tube. Thus, yield is sacrificed for convenience.

Coupled Transcription: Translation

Unlike eukaryotic systems where transcription and translation occur sequentially, in *E. coli*, transcription and translation occur simultaneously within the cell. In vitro *E. coli* translation systems are thus performed the same way, coupled, in the same tube under the same reaction conditions (one-step reaction). During transcription, the 5' end of the RNA becomes available for ribosomal binding and undergoes translation while its 3' end is still being transcribed. This early binding of ribosomes to the RNA maintains transcript stability and promotes efficient translation. This bacterial translation system gives efficient expression of either prokaryotic or eukaryotic gene products in a short amount of time. For the highest protein yield and the best initiation fidelity, make sure the DNA template has a Shine-Dalgarno ribosome binding site upstream of the initiator codon. Capping of eukaryotic RNA is not required. Use of *E.coli* extract also eliminates cross-reactivity or other problems associated with endogenous proteins in eukaryotic lysates. Also, the *E. coli* S30 extract system allows expression from DNA vectors containing natural *E. coli* promoter sequences (such as *lac* or *tac*).

Important Elements for Translation

There are some significant differences between prokaryotic and eukaryotic mRNA transcripts. Typically, eukaryotic mRNAs are characterized by two post-transcriptional modifications: a 5'-7 methyl-GTP cap and a 3' poly (A) tail. Both modifications contribute to the stability of the mRNA by preventing degradation. Additionally, the 5' cap structure enhances the translation of mRNA by helping to bind the eukaryotic ribosome and assuring recognition of the proper AUG initiator codon. This function may vary with the translation system and with the specific mRNA being synthesized. The consensus sequence 5'-GCCACCAUGG-3', also known as the "Kozak" sequence, is considered to be the strongest ribosomal binding signal in eukaryotic mRNA. For efficient translation initiation, the key elements are the G residue at the +1 position and the A residue at the -3 position. An mRNA that lacks the Kozak consensus sequence may be translated efficiently in eukaryotic cell-free systems if it possesses a moderately long 5'-untranslated region (UTR) that lacks stable secondary structure.

In bacteria, the ribosome is guided to the AUG initiation site by a purine-rich region called the Shine-Dalgarno (SD) sequence. This sequence is complementary to the 3' end of the 16s rRNA in the 30S ribosomal subunit. Upstream from the initiation AUG codon, the SD region has the consensus sequence 5'-UAAGGAGGUGA-3'. Specific mRNAs vary considerably in the number of nucleotides that complement the anti-Shine-Dalgarno sequence of 16S rRNA, ranging from as few as two to nine or more. The position of the ribosome binding site (RBS) in relation to the AUG initiator is very important for efficiency of translation (usually from -6 to -10 relative to the A of the initiation site).

DIFFERENT EXPRESSION SYSTEMS FOR RECOMBINANT PROTEINS

Expression system

The expression of high levels of stable and functional proteins remains a bottleneck in many scientific endeavors, including the determination of structures in a high-throughput fashion or the screening for novel active compounds in modern drug discovery. Recently, numerous developments have been made to improve the production of soluble and active proteins in heterologous expression systems. These include modifications to the expression constructs, the introduction of new and/or improved pro- and eukaryotic expression systems, and the development of improved cell-free protein synthesis systems. The introduction of robotics has enabled a massive parallelization of expression experiments, thereby vastly increasing the throughput and, hopefully, the output of such experiments.

Generally, protein expression systems can be categorized into the following groups:

1) Prokaryotic expression systems.
2) Yeast expression system.
3) Insect expression system.
4) Mammalian cells.

Prokaryotic Expression System

Microorganisms like the enterobacterium *Escherichia coli* are outstanding factories for recombinant expression of proteins. An expression system for the production of recombinant proteins in *E. coli* usually involves a combination of a plasmid and a strain of *E. coli*. The main purpose of recombinant protein expression is often to obtain a high degree of accumulation of soluble product in the bacterial cell. This strategy is not always accepted by the metabolic system of the host and in some situations a cellular stress response is encountered. Another response encountered in recombinant systems is the accumulation of target proteins into insoluble aggregates known as inclusion bodies. These aggregated proteins are in general misfolded and thus biologically inactive.

Under normal cellular conditions a subset of cytoplasmic proteins are able to fold spontaneously while aggregation prone proteins require the existence of a number of molecular chaperones that interact reversibly with nascent polypeptide chains to prevent aggregation during the folding process. Aggregation of recombinant proteins overexpressed in bacterial cells could therefore result either from accumulation of high concentrations of folding intermediates or from inefficient processing by molecular chaperones. No universal approach has been established for the efficient folding of aggregation prone recombinant proteins.

The literature describes a number of methods for the redirection of proteins from inclusion bodies into the soluble cytoplasmic fraction. Overall they can be divided into procedures where protein is refolded from inclusion bodies and procedures where the expression strategy is modified to obtain soluble expression. Refolding from inclusion bodies is in many cases considered undesireable, but is however sometimes the method of choice. The major obstacles are the poor recovery yields, the requirement for optimization of refolding conditions for each target protein and the possibility that the resolubilization procedures could affect the integrity of refolded proteins. In addition, the purification of highly expressed soluble protein is less expensive and time consuming than refolding and purification from inclusion bodies. Maximizing the production of recombinant proteins in a soluble form is therefore an attractive alternative to *in vitro* refolding procedures. The methods used to mediate soluble expression can be divided into procedures where target modification is avoided and procedures where the target sequence is engineered.

Strategies where target modification is avoided

Some proteins directly influence the cellular metabolism of the host by their catalytic properties, but in general expression of recombinant proteins induces a "metabolic burden". The metabolic burden is defined as the amount of resources (raw material and energy), which are withdrawn from the host metabolism for maintenance and expression of the foreign DNA. The formation of inclusion bodies occurs as a response to the accumulation of denatured protein. The metabolic burden and inclusion body formation are not directly linked but are both among the main factors to determine the ability of cells to produce soluble recombinant protein. Since the accumulation of denatured protein and the metabolic burden can be controlled by a number of environmental factors, we are partially able to control the formation of soluble protein *in vivo*.

Protein expression at reduced temperatures

A well known technique to limit the *in vivo* aggregation of recombinant proteins consists of cultivation at reduced temperatures. This strategy has proven effective in improving the solubility of a number of difficult proteins including human interferon α-2, subtilisin E, ricin A chain, bacterial luciferase, Fab fragments, β-lactamase, rice lipoxygenase L-2, soybean lypoxygenase L-1, kanamycin nuclotidyltransferase and rabbit muscle glycogen phosphorylase. The aggregation reaction is in general favored at higher temperatures due to the strong temperature dependence of hydrophobic interactions that determine the aggregation reaction. A direct consequence of temperature reduction is the partial elimination of heat shock proteases that are induced under over expression conditions. Furthermore, the activity and expression of a number of *E. coli* chaperones are increased at temperatures around 30°C. The increased stability and potential for correct folding at low temperatures are partially explained by these factors.

However, a sudden decrease in cultivation temperature inhibits replication, transcription and translation. Traditional promoters used in vectors for recombinant protein expression are also strongly affected in terms of efficiency. A similar transcriptional effect is achieved when a moderately strong or weak promoter is used or when a strong promoter is partially induced. Low induction levels have been found to result in higher amounts of soluble protein. This is a result of the reduction in cellular protein concentration which favors folding. However, bacterial growth is decreased, thus resulting in a decreased amount of biomass.

Modification of cultivation strategies to obtain soluble protein

The simplest way to produce a recombinant protein is by batch cultivation. Here all nutrients required for growth are supplied from the beginning and there is a limited control of the growth during the process. This limitation often leads to changes in the growth medium such as changes in pH and concentration of dissolved oxygen as well as substrate depletion. Furthermore inhibitory products of various metabolic pathways accumulate. Cell densities and production levels are only moderate in batch cultivations.

In fed batch cultivations, the concentration of energy sources can be adjusted according to the rate of consumption. Several other factors can also be regulated in order to obtain the maximal production level in terms of target protein per biomass. The formation of inclusion bodies can be followed in fed batch cultivations by monitoring changes in intrinsic light scattering by flow cytometry. This allows for real time optimization of growth conditions as soon as inclusion bodies are detected even at low levels and inclusion body formation can potentially be avoided.

Folding of some proteins require the existence of a specific cofactor. Addition of such cofactors or binding partners to the cultivation media may increase the yield of soluble protein dramatically. This was demonstrated for a recombinant mutant of hemoglobin for which the accumulation of soluble product was improved when heme was in excess. Similarly, a 50% increase in solubility was observed for gloshedobin when *E. coli* recombinants were cultivated in the presence of 0.1 mM Mg^{2+}. An important factor in soluble expression of recombinant proteins is media composition and optimization. Although this is attained mostly by trial and error, it nevertheless may be beneficial.

Molecular chaperones drive folding of recombinant proteins

A possible strategy for the prevention of inclusion body formation is the co-overexpression of molecular chaperones. This strategy is attractive but there is no guarantee that chaperones improve recombinant protein solubility. *E. coli* encode chaperones, some of which drive folding attempts, whereas others prevent protein aggregation. As soon as newly synthesized proteins leave the exit tunnel of the *E. coli* ribosome they associate with the trigger factor chaperone. Exposed hydrophobic patches on newly synthesized proteins are protected by association with trigger factor from unintended inter- or intramolecular interactions thus preventing premature folding. Proteins can start or continue their folding into the native state after release from trigger factor. Proteins trapped in non-native and aggregation prone conformations, are substrates for DnaK and GroEL. DnaK (Hsp70 chaperone family) prevents the formation of inclusion bodies by reducing aggregation and promoting proteolysis of misfolded proteins. A bi-chaperone system involving DnaK and ClpB (Hsp100 chaperone family) mediates the solubilization or disaggregation of proteins. GroEL (Hsp60 chaperone family) operates the protein transit between soluble and insoluble protein fractions and participates positively in disaggregation and inclusion body formation. Small heat shock proteins lbpA and lbpB protect heat denatured proteins from irreversible aggregation and have been found associated with inclusion bodies.

Simultaneous over-expression of chaperone encoding genes and recombinant target proteins proved effective in several instances. Co-overexpression of trigger factor in recombinants prevented the aggregation of mouse endostatin, human oxygen-regulated protein ORP150, human lysozyme and guinea pig liver transglutaminase. Soluble expression was further stimulated by the co-overexpression of the GroEL-GroES and DnaK-DnaJ-GrpE chaperone systems along with trigger factor. The chaperone systems are cooperative and the most favorable strategies involve co-expression of combinations of chaperones belonging to the GroEL, DnaK, ClpB and ribosome associated trigger factor families of chaperones.

Interaction partners and protein folding

Protein insolubility in the *E. coli* cytoplasm is partially related to the distribution of hydrophobic residues on the surface of the protein. The soluble expression of subunits of hetero multimeric proteins therefore sometimes suffers from inclusion body formation in the absence of an appropriate binding partner. Soluble expression in *E. coli* of the bacteriophage T4 gene 23 product (major capsid protein) required the co-expression of gene product 31 (phage co-chaperonin gp31). Expression of the correct interaction partner enabled gp23 to fold correctly and form long regular structures in the cytoplasm of *E. coli*. Another study reports the purification of a heterodimeric complex by expression of each subunit (pheromaxein A and C) as a fusion to thioredoxin. Each subunit remained soluble in solution, when thioredoxin was proteolytically removed, only in the presence of the other.

Conclusively, interaction partners potentially favour *in vivo* solubility of target proteins. New systems for co-expression of multiple proteins involved in complex structures enable such strategies.

Eukaryotic Expression System

Yeast is a eukaryotic organism and has some advantages and disadvantages for protein expression as compared to *E. coli*. One of the major advantages is that yeast cultures can be grown to very high densities, which makes them especially useful for the production of isotope labeled protein for NMR. The yeast species, *Saccharomyces cerevisiae* has proven to be extremely useful for expression and analysis of eukaryotic proteins. This yeast strain has been genetically well characterized and are known to perform many posttranslational modifications. These single-celled eukaryotic organisms grow quickly in defined medium, are easier and less expensive to work with than insect or mammalian cells, and are easily adapted to fermentation. Yeast expression systems are ideally suited for large-scale production of recombinant eukaryotic proteins. In some instances the most cost-effective expression of functional enzymes is the yeast expression system. Yeast is an established industrial fermentation system and supports high-level recombinant protein production. The major advantages of yeast expression system are:

1) High yield
2) High productivity
3) Chemically defined media
4) Product processing similar to mammalian cells
5) Stable production strains
6) Durability
7) Lower protein production cost

High cell densities

When yeast is grown with the high-cell-density fermentation technology, unprecedented levels of cell mass per liter of fermentation fluid are produced. The system has attained dry-cell-weight densities exceeding 100 gram/liter and continuous fermentation productivities of 10 to 12 grams of recombinant protein/liter/hour.

Controllable process

The growth medium that feeds yeast is completely defined. It consists of a simple, inexpensive formulation. The carbon source is fed to the fermentor at a rate designed to achieve maximum cell density while maintaining optimal production of foreign protein. This process minimizes any toxic effects the foreign protein might have on the yeast.

Mammalian-like proteins

As a eukaryotic system, the Yeast Expression System produces mammalian-like proteins. For example, the expression of Hepatitis B surface antigen (HBsAg) in yeast leads to production of particles that are immunoreactive with anti-HBsAg antibodies. These particles are similar to Dane particles isolated from the sera of human carriers.

Generations of stability

Expression of foreign genes is achieved by integration of foreign DNA into the chromosomal DNA of the host genome. The integrated DNA is stable for many generations; all cells can produce the protein. In contrast, plasmid-based systems require selective pressure on plasmids to maintain the foreign DNA. Cells that lose the plasmid cannot produce the desired foreign protein.

Durability

The Yeast Expression System requires no special handling. It was developed to withstand the adverse conditions of large scale, continuous fermentors. This feature makes yeast able to survive unexpected disruptions in the fermentation process.

Maximum Value

High per-cell expression levels combined with high cell-density growth of yeast translates into greater quantities of recombinant protein per fermentor volume. This reduces production costs by increasing the amount of product per fermentation run. Protein purification is another cost-saving area. The yeast system can secrete protein into the medium, so the broth that enters purification contains a higher concentration of the desired protein. Pure protein is recovered with higher yield and lower cost.

Insect Cell Expression System

Insect cells are a higher eukaryotic system than yeast and are able to carry out more complex post-translational modifications than the other two systems. They also have the best machinery for the folding of mammalian proteins and, therefore, give you the best chance of obtaining soluble protein when you want to express a protein of mammalian origin. The most commonly used vector system for recombinant protein expression in insect is baculovirus, although baculoviral also can be used for gene transfer and expression in mammalian cells.

Advantages

Baculovirus-assisted insect cell expression is optimal for glycosylated protein expression in a cost-effective manner. There are many advantages to using baculovirus for heterologous gene expression. Heterologous cDNA is expressed well. Proper transcriptional processing of genes with introns occurs but is expressed less efficiently. As with other eukaryotic expression systems, baculovirus expression of heterologous genes permits folding, post-translational modification and oligomerization in manners that are often identical to those that occur in mammalian cells. The insect cytoplasmic environment allows proper folding and S-S bond formation, unlike the reducing environment of the *E. coli* cytoplasm. Post-translational processing identical to that of mammalian cells has been reported for many proteins. These include proper proteolysis, N- and O-glycosylation, acylation, amidation, carboxymethylation, phosphorylation, and prenylation. Proteins may be secreted from cells or targeted to different subcellular locations. Single polypeptide, dimeric and trimeric proteins have been expressed in baculoviruses. Finally, expression of heterologous proteins is under the control of the strong polyhedrin promoter, allowing levels of expression of up to 30% of the total cell protein.

The benefits of protein expression with baculovirus can be summarized as:

1) Eukaryotic post-translational modification
2) Proper protein folding and function
3) High expression levels
4) Easy scale up with high-density suspension culture
5) Safety

Baculoviruses infect primarily insects with a narrow host range. *Autographa californica*, the most commonly used baculovirus for protein expression, infects only 2 lepidopteran (moth) families in nature. Although these viruses may enter other cells types (perhaps by phagocytosis), they are not infectious in them. For example, nucleocapsid proteins are not removed in most human cells. In human hepatic cell lines that do remove these proteins, the virus fails to replicate and express proteins due to the absence of insect transcription factors. Thus, working with baculoviruses is considered safe for humans and contamination of mammalian cell lines in shared biosafety hoods is not a problem. Recombinant DNA guidelines recommend a BL1 biosafety level for most baculovirus expression experiments.

Disadvantages

Despite these potential advantages, particular patterns of post-translational processing and expression must be empirically determined for each construct. Differences in proteins expressed by mammalian and baculovirus infected insect cells have been described and overcome in some cases. For example, inefficient secretion from insect cells may be circumvented by the addition of insect secretion signals (ex. honeybee melittin sequence). Improperly folded proteins and proteins that occur as intracellular aggregates may be due to expression late in the infection cycle. In such cases, harvesting cells at earlier times after infection may help. Low levels of expression can often be increased with optimization of time of expression and multiplicity of infection. The complete analysis of carbohydrate structures has been reported for a limited number of glycoproteins. Potential N-linked glycosylation sites are often either fully glycosylated or not glycosylated at all, as opposed to expression of various glycoforms that may occur in mammalian cells. Species-specific or tissue-specific modifications are unlikely to occur.

Mammalian Expression System

The production of proteins in mammalian cells is an important tool in numerous scientific and commercial areas. For example, the proteins expressed in and purified from mammalian cell system are routinely needed for life science research and development. In the field of biomedicine, proteins for human therapy, vaccination or diagnostic applications are typically produced in mammalian cells. Gene cloning, protein engineering, biochemical and biophysical characterization of proteins also require the use of gene expression in mammalian cells. Other applications in widespread use involve screening of libraries of chemical compounds in drug discovery, and the development of cell-based biosensors.

Usage of mammalian expression

The proteins produced in the mammalian expression system have the best structural and functional features that are usually most close to their cognate native form and can satisfy the following application needs or utility:

1) Transgene expression
2) Biochemical analyses
3) Assay standards
4) Functional studies of the protein (in vitro and ex vivo)
5) Structural studies, including protein crystallization, protein structure and NMR
6) Protein-protein interaction experiments

7) Enzyme kinetics
8) Immunogen for antibodies development
9) Proteomic and phenomics study
10) Drug target discovery and validation
11) Cell line development, drug screening, and in vitro model system
12) Animal studies, including in vivo functional and ADME, PK/TK and safety studies
13) Physiology and pathology studies
14) Diagnostic application
15) Therapeutic application
16) Prophylactic (vaccine) development
17) Protein engineering and mutagenesis studies

PRODUCTION OF FUSION PROTEINS

This might at first seem a disadvantage because the natural product of the inserted gene is not made. However, the extra amino acids on the fusion protein can be a great help in purifying the protein product. Oligo-histidine is one of the commonly used fusion tags for protein expression (in both prokaryotic and eukaryotic system). Oligo-histidine regions like this have a high affinity for metals like nickel, so the expressed target proteins can be purified using nickel affinity chromatography. The beauty of this method is its simplicity and speed. After the expression systems have made the fusion protein, it can be applied to a nickel affinity column, wash out all unbound proteins, and then release the fusion protein with histidine or a histidine analog called imidazole. This procedure allows harvesting essentially pure fusion in only one step. This is possible because very few if any natural proteins have Oligo-histidine regions, so the fusion protein is essentially the only one that binds to the column. There are other tags have been used in protein expression in mammalian cells. There are mechanisms designed for removing tag from the fusion proteins. For example, enterokinase have been used for cleave tags from the fusion proteins that contain the cleave site for the enzymes.

Protein expression and purification have neither a fixed approach nor a protocol for all. Every project of expressing and purifying a protein is unique and requires specific analyses, from bioinformatics analysis and engineering design of protein sequence, through selection of expression vector and host system, to meeting various requirements or purposes of the final product (expressed and purified protein). Compared with recombinant DNA technologies, protein expression and purification technologies are much more complicated and sophisticated. This is why there are so many methods and protocols developed for successfully carrying out a recombinant protein project.

Insulin

Since Banting and Best discovered the hormone, insulin in 1921, diabetic patients, whose elevated sugar levels are due to impaired insulin production, have been treated with insulin derived from the pancreas glands of abattoir animals. The hormone, produced and secreted by the beta cells of the pancreas' islets of Langerhans, regulates the use and storage of food, particularly carbohydrates. Although bovine and porcine insulin are similar to human insulin, their composition is slightly different. Consequently, a number of patients' immune systems produce antibodies against it, neutralising its actions and resulting in inflammatory responses at injection sites. Added to these adverse effects of bovine and porcine insulin, were fears of long term complications ensuing from the regular injection of a foreign substance, as well as a projected decline in the production of animal derived insulin. These factors led researchers to consider synthesising *Humulin* by inserting the insulin gene into a suitable vector, the *E. coli* bacterial cell, to produce an insulin that is chemically identical to its naturally produced counterpart. This has been achieved using Recombinant DNA technology. This method is a more reliable and sustainable method than extracting and purifying the abattoir by-product.

The structure of insulin

Chemically, insulin is a small, simple protein. It consists of 51 amino acid, 30 of which constitute one polypeptide chain, and 21 of which comprise a second chain. The two chains are linked by a disulfide bond. The genetic code for insulin is found in the DNA at the top of the short arm of the eleventh chromosome. It contains 153 nitrogen bases (63 in the A chain and 90 in the B chain). DNA (Deoxyribolnucleic Acid), which makes up the chromosome, consists of two long intertwined helices, constructed from a chain of nucleotides, each composed of a sugar deoxyribose, a phosphate and nitrogen base. There are four different nitrogen bases, adenine, thymine, cytosine and guanine. The synthesis of a particular protein such as insulin is determined by the sequence in which these bases are repeated.

A weakened strain of the common bacterium, *Escherrichia coli* (*E. coli*), an inhabitant of the human digestive tract, is the 'factory' used in the genetic engineering of insulin. When the bacterium reproduces, the insulin gene is replicated along with the plasmid, a circular section of DNA. *E. coli* produces enzymes that rapidly degrade foreign proteins such as insulin. By using mutant strains that lack these enzymes, the problem is avoided. In *E. coli*, β -galactosidase is the enzyme that controls the transcription of the genes. To make the bacteria produce insulin, the insulin gene needs to be tied to this enzyme.

Manufacturing Humulin

The first step is to chemically synthesise the DNA chains that carry the specific nucleotide sequences characterising the A and B polypeptide chains of insulin. The required DNA sequence can be determined because the amino acid compositions of both chains have been charted. Sixty three nucleotides are required for synthesising the A chain and ninety for the B chain, plus a codon at the end of each chain,signalling the termination of protein synthesis. An anti-codon, incorporating the amino acid, methionine, is then placed at the beginning of each chain which allows the removal of the insulin protein from the bacterial cell's amino acids. The synthetic A and B chain 'genes' are then separately inserted into the gene for a bacterial enzyme, β-galactosidase, which is carried in the vector's plasmid. At this stage, it is crucial to ensure that the codons of the synthetic gene are compatible with those of the β-galactosidase.

The recombinant plasmids are then introduced into *E. coli* cells. Practical use of Recombinant DNA technology in the synthesis of human insulin requires millions of copies of the bacteria whose plasmid has been combined with the insulin gene in order to yield insulin. The insulin gene is expressed as it replicates with the β-galactosidase in the cell undergoing mitosis. The protein which is formed, consists partly of β -galactosidase, joined to either the A or B chain of insulin. The A and B chains are then extracted from the β - galactosidase fragment and purified. The two chains are mixed and reconnected in a reaction that forms the disulfide cross bridges, resulting in pure *Humulin* - synthetic human insulin.

Biological implications of genetically engineered Recombinant human insulin

Human insulin is the only animal protein to have been made in bacteria in such a way that its structure is absolutely identical to that of the natural molecule. This reduces the possibility of complications resulting from antibody production. In chemical and pharmacological studies, commercially available Recombinant DNA human insulin has proven indistinguishable from pancreatic human insulin. Initially the major difficulty encountered was the contamination of the final product by the host cells, increasing the risk of contamination in the fermentation broth. This danger was eradicated by the introduction of purification processes. When the final insulin product is subjected to a battery of tests, including the finest radio-immuno assay techniques, no impurities can be detected. The entire procedure is now performed using yeast cells as a growth medium, as they secrete an almost complete human insulin molecule with perfect three dimensional structure. This minimises the need for complex and costly purification procedures.

Although the production of human insulin is unarguable welcomed by the majority of insulin dependent patients, the existence of a minority of diabetics who are unhappy with the product cannot be ignored. Although not a new drug, the insulin derived from this new method of production must continue to be studied and evaluated, to ensure that all its users have the opportunity to enjoy a complication free existence.

hGH Production by Using Molecular Biotechnology Techniques

Human Growth Hormone (HGH or hGH) is the most abundant hormone produced by the pituitary gland (pituitary is one of the endocrine glands). The pituitary gland is located in the center of the brain. HGH is also a very complex hormone. It is made up of 191 amino acids - making it fairly large for a hormone. In fact, it is the largest protein created by the Pituitary gland. HGH secretion reaches its peak in the body during adolescence. This makes sense because HGH helps stimulate our body to grow. But, HGH secretion does not stop after adolescence. Our body continues to produce HGH usually in short bursts during deep sleep. Growth Hormone is known to be critical for tissue repair, muscle growth, healing, brain function, physical and mental health, bone strength, energy and metabolism. In short, it is very important to just about every aspect of our life.

Children with hypopituitary dwarfism do not grow in size in normal proportions like children with normal pituitary function because their bodies make too little growth hormone. They are destined to be dwarfs unless they are treated with hGH. Growth hormone from other animals may not help. Moreover, early preparations of hGH were contaminated with pyrogens or other contaminants. Growth hormone is a 191 amino acid protein that is produced in the pituitary gland and regulates growth and development. The production of HGH was achieved by constructing a hybrid gene from the natural hGH cDNA and synthetic oligonucleotides that encode the amino terminus of the mature form of the protein. This coding sequence was ligated into a plasmid adjacent to a bacterial promoter.

Human growth hormone is produced by the bacteria and then secreted with the concomitant removal of the signal peptide by bacterial proteases. The only difference between the secreted hGH and that produced intracellularly is the presence of an amino-terminal methionine on the intracellular expressed molecule. Because the secreted form lacks this methionine, it is called met-less hGH. Prior to the use of recombinant DNA technology to modify bacteria to produce human growth hormone, the hormone was manufactured by extraction from the pituitary glands of cadavers, as animal growth hormones have no therapeutic value in humans. Production of a single year's supply of human growth hormone required up to fifty pituitary glands, creating significant shortages of the hormone.In 1979, scientists at Genentech produced human growth hormone by inserting DNA coding for human growth hormone into a plasmid that was implanted in escherichia coli bacteria. The gene that was inserted into the plasmid was created by reverse transcription of the mRNA found in pituitary glands to complementary DNA. HaeIII, a type of restriction enzyme which acts at restriction sites "in the 3' noncoding region" and at the 23rd codon in complementary DNA for human growth hormone, was used to produce "a DNA fragment of 551 base pairs which includes coding sequences for amino acids 24–191 of hGH. Then a chemically synthesized DNA 'adaptor' fragment containing an ATG initiation codon..."was produced with the codons for the first through 23rd amino acids in human growth hormone. The "two DNA fragments... [Were] combined to form a synthetic-natural 'hybrid' gene." The use of entirely synthetic methods of DNA production to produce a gene that would be translated to human growth hormone in *Escherichia coli* would have been exceedingly laborious due to the significant length of the amino acid sequence in human growth hormone. However, if the cDNA reverse transcribed from the mRNA for human growth hormone were inserted directly into the plasmid inserted into the *Escherichia coli*, the bacteria would translate regions of the gene that are not translated in humans, thereby producing a "pre-hormone containing an extra 26 amino acids" which might be difficult to remove.

Tissue Plasminogen Activator

Plasmin is a protease which catalyses the proteolytic degradation of fibrin present in clots, thus effectively dissolving the clot. Plasmin is derived from plasminogen, its circulating zymogen. Plasminogen is synthesized in and released from the kidneys. It is a single-chain 90-kDa glycoprotein, which is stabilized by several disulphide linkages. Tissue plasminogen activator (tPA, also known as fibrinokinase) represents the most important physiological activator of plasminogen. tPA is a 527 -amino acid serine protease. It is synthesized predominantly in vascular endothelial cells (cells lining the inside of blood vessels). tPA displays four potential glycosylation sites, three of which are normally glycosylated (residues 117, 184 and 448). It is normally found in the blood in two forms: a single-chain polypeptide (type I tPA) and a two-chain structure (type II) proteolytically derived from the single-chain structure. The two-chain form is the one predominantly associated with clots undergoing lysis, but both forms display fibrinolytic activity.

Fibrin contains binding sites for both plasminogen and tPA, thus bringing these into close proximity. This facilitates direct activation of the plasminogen at the clot surface. This activation process is potentiated by the fact that binding of tPA to fibrin (a) enhances the subsequent binding of plasminogen and (b) increases tPA's activity towards plasminogen by up to 600-fold. Overall therefore, activation of the thrombolytic cascade occurs exactly where it is needed on the surface of the clot. This is important as the substrate specificity of plasmin is poor, and circulating plasmin displays the catalytic potential to proteolyse fibrinogen, factor V and factor VIII. Although soluble serum tPA displays a much reduced activity towards plasminogen, some freely circulating plasmin is produced by this reaction. If uncontrolled, this could increase the risk of subsequent haemorrhage. This scenario is usually averted as circulating plasmin is rapidly neutralized by another plasma protein, a2 -antiplasmin. (a2 -antiplasmin, a 70-kDa, single-chain glycoprotein, binds plasmin very tightly in a 1: 1 complex).

In contrast to free plasmin, plasmin present on a clot surface is very slowly inactivated by a1 -antiplasmin. The thrombolytic system has thus evolved in a self-regulating fashion, which facilitates efficient clot degradation with minimal potential disruption to other elements of the haemostatic mechanism. Production of tPA Although tPA was first studied in the late 1940s, its extensive characterization was hampered by the low levels at which it is normally synthesized. Detailed studies were facilitated in the 1980s after the discovery that the Bowes melanoma cell line produces and secretes large quantities of this protein. This also facilitated its initial clinical appraisal. The tPA gene was cloned from the melanoma cell line in 1983, this facilitated subsequent large-scale production in CHO cell lines by recombinant DNA technology. The tPA cDNA contains 2530 nucleotides and encodes a mature protein of 527 amino acids. The glycosylation pattern was similar, though not identical, to the native human molecule. A marketing license for the product as first issued in the USA to Genentech in 1987 (under the trade name Alteplase).

The therapeutic indication was for the treatment of acute myocardial infarction. The production process entails a initial (10,000 litre) formation step, during which the cultured CHO cells produce and secrete tPA into the fermentation medium. After removal of the cells by submicron filtration and initial concentration, the product is purified by a combination of several chromatographic steps. The final product has been shown to be greater then 99% pure by several analytical techniques, including HPLC, SDS-PAGE, tryptic mapping and N-terminal sequencing. Alteplase has proved to be effective in the early treatment of patients with acute myocardial infarction (i.e., those treated within 12 hours after the first symptoms occur), and has significantly increased rates of patient survival (as measured one day and 30 days after the initial event). tPA has thus established itself as a first line option in the management of acute myocardial infarction. A therapeutic dose of 90-100 mg (often administered by infusion over 90 minutes) results in a steady-state Alteplase concentration of 3-4 mg/litre during that period. The product is, however, cleared rapidly by the liver, displaying a serum half-life of minutes. As is the case for most thrombolytic agents, the most significant risk associated with tPA administration is the possible induction of severe haemorrhage.

Modified forms of tPA have also been generated in an effort to develop a product with an improved therapeutic profile (e.g. faster acting or exhibiting a prolonged plasma half-life). Ecokinase is the trade name given to one such modified human tPA produced in recombinant *E. coli* cells. It gained marketing approval in Europe in 1996. Its development was based upon the generation of a synthetic nucleotide sequence encoding a shortened (355-amino acid) tPA molecule. This analogue contained only the tPA domains responsible for fibrin selectivity and catalytic activity. The nucleotide sequence was integrated into an expression vector subsequently introduced into E. coli (strain KI2), by treatment with calcium chloride. The protein is expressed intracellularly, where it accumulates in the form of an inclusion body. Due to the prokaryotic production system, the product is non-glycosylated. The final sterile freeze-dried product is equally as effective as Alteplase and exhibits a two-year shelf life when stored at temperatures below 25°C.

Clotting Factor

An example of a recombinant pharmaceutical produced in eukaryotic cells in human factor VIII, a protein which plays a central role in blood clotting. The commonest form of haemophilia in humans results from an inability to synthesize factor VIII, leading to a breakdown in the blood clotting pathway and the well-known symptoms associated with the disease. The only way to treat haemophilia is by injection of purified factor VIII protein, obtained from human blood provided by the donors. Purification of factor VIII is a complex procedure and the treatment is very expensive. Recombinant factor VIII, free from contamination problems would be a significant achievement for biotechnology. Factor VIII gene is very large, over 186 kb in length, and is split into 26 exons and 25 introns. The mRNA codes for a large polypeptide (2351 amino acids) which undergoes a complex series of post-translational processing events, eventually resulting in a dimeric protein consisting of a large subunit, derived from the upstream region of the initial polypeptide, and a small subunit from the downstream segment. It has not been possible to synthesize an active version in *E. coli.*

The synthesis and purification of proteins from cloned genes is one of the most important aspects of genetic manipulation, particularly where valuable therapeutic proteins are concerned.

Interferon

Many types of human and animal cells produce, as a consequence of exposure to certain viruses or other inducing agents, glycoproteins known as interferons, which have a molecular weight of about 20,000. Interest in interferons first arose from their ability to render cells resistant to virus attack, and subsequently from the possibility that interferons may be useful as anticancer agents. However, supplies of interferons are severely limited; for example, two litres of human blood (one of the principle sources) are required to produce 1 mg of purified human leucocyte interferon. This made interferon an obvious target for gene cloning. Goeddel et al. (1980) obtained a recombinant plasmid with an insert of about 1000 bp, which could be excised and cloned into another vector to place it under the control of a trp promoter. This gave a much higher level of expression (about 480,000 units per litre of culture). Knowledge of the DNA sequence of the 1000-bp fragment enabled a further series of manipulations to be carried out to obtain expression of mature leucocyte interferon at a high level. This involved removing the sequence coding for the signal precursor initiation ATG codon. Another DNA fragment was then inserted which carried the *E. coli* trp operator/promoter region and ribosome-binding site. One clone obtained after these manipulations produced 2.5 x 108 units of interferons per litre of culture. The interferon produced was shown to have biological activity by its ability to protect squirrel monkeys against encephalomyocarditis virus.

Recombinant Vaccines

A recombinant vaccine contains either a protein or a gene encoding a protein of a pathogen origin that is immunogenic and critical to the pathogen function; the vaccine is produced using recombinant DNA technology. The vaccines based on recombinant proteins are also called subunit vaccines. The logic of such vaccines, in simple terms, is as follows. Proteins are generally immunogenic, and many of them are critical for the pathogenic organism. The genes encoding such proteins can be identified and isolated from a pathogen and expressed in *E. coli* or some other suitable host for a mass production of the proteins.

The concerned proteins are then purified and mixed with suitable stabilizers and adjuvant, if required, and used for immunization. The different steps involved in the development of a recombinant protein based vaccine may be simply summarised as follows:

(i) The first step is to identify a protein that is both immunogenic and critical for the pathogen.

(ii) The gene encoding this protein is then identified and isolated.

(iii) The gene is integrated into a suitable expression vector and introduced into a suitable host where it expresses the protein in large quantities.

(iv) The protein is then isolated and purified from the host cells.

(v) It is used for the preparation of vaccine. The host organisms used for expression of immunogenic proteins to be used as vaccines may be anyone of the following.

1) A genetically engineered microorganism, e.g., yeast for the expression of hepatitis B surface antigen (HBsAg) used as vaccine against hepatitis B virus (approved for marketing in India).

2) Cultured animal cells, e.g., HBsAg expressed in CHO (Chinese hamster ovary) cell line and C-127 cell line.

3) Transgenic plants, e.g., HBsAg, HIV-l (human immunodeficiency virus-I) epitope (in experimental stages).

4) Insect larvae; the gene is integrated into a bacculovirus genome, which is used to infect insect larvae. Often a very high quantity of the recombinant protein is produced.

So far a large number of recombinant immunogenic proteins of pathogens have been produced and evaluated. In general, a majority of such proteins are ineffective or only poorly effective in immunization.

Plant Derived Vaccines

Conventional vaccines consist of attenuated or inactivated pathogens. In case of many pathogens, the gene encoding a critical antigen has been isolated and expressed in bacteria/animals, and the recombinant protein so produced is used as a vaccine; such vaccines are called recombinant vaccines. Recombinant vaccines are produced through bacterial fermation or in animal cell cultures, which often makes their cost prohibitively high. In addition, storage and transport of vaccines, especially in developing countries, presents many problems, e.g., cold storage.

Therefore, plants are being developed as an alternative vaccine production and delivery systems. There are two basic objectives of such efforts:

1) Development of edible vaccines.

2) Production of recombinant antigenic proteins to be used as vaccines.

Recombinant Vaccines For Animal Use

Several commercially used recombinant vaccines used on animals employ a vector based delivery system. These include the VRG vaccine, which protects animals against rabies, and the Purevax recombinant feline leukaemia vaccine. As mentioned above,the VRG vaccine consists of a recombinant *Vaccinia* virus that carries the gene for a rabies glycoprotein.The virus has been modified in several ways, one of which involves the removal of the thymidine kinase gene (Fig. 11.2), making it safer to administer than in its original form. Studies have in fact shown it has not caused any side effects in over 10 avian species and 35 mammalian species. The Purevax leukaemia vaccine contains a harmless recombinant canarypox virus that incorporates the FeLV gene. This gene produces a protein identical to that produced by the FeLV (feline leukaemia) virus, with the result that the cat's immune response is triggered without the danger of the actual virus being introduced. The canarypox virus is also used as a vector in dog and ferret vaccines.

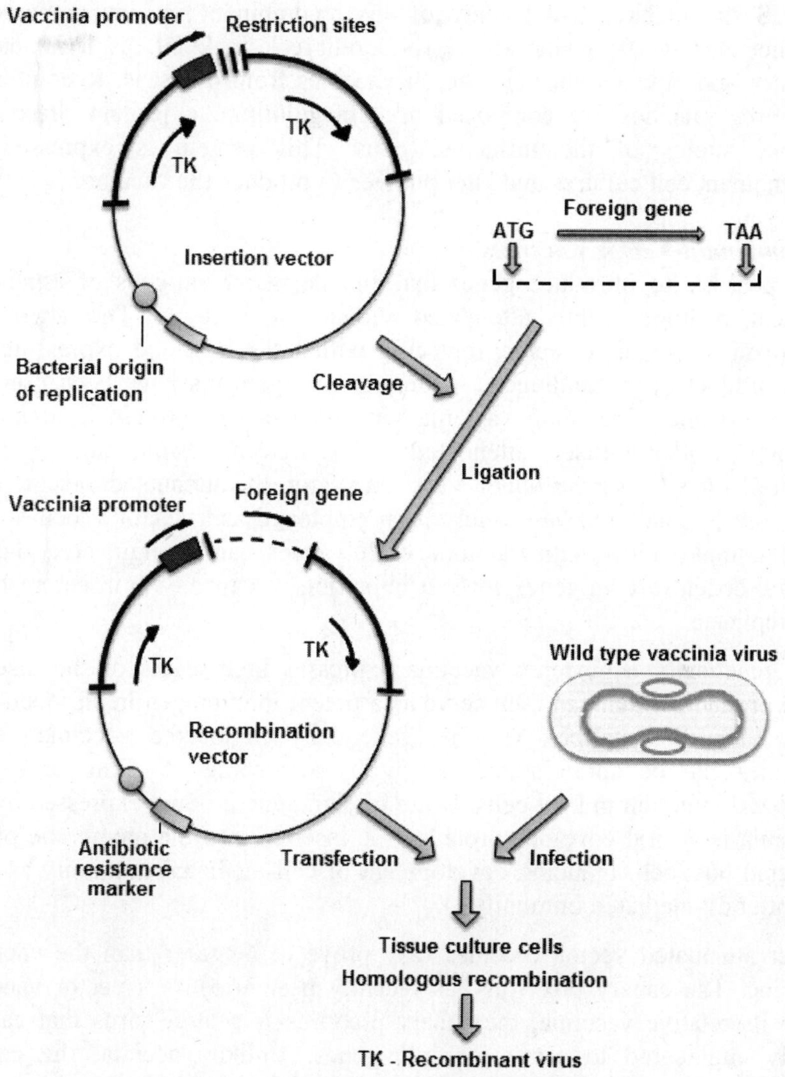

Fig. 11.2: Recombinant Vaccine

Recombinant Vaccines Used by Humans

Besides Gardisal cervical cancer vaccine, one of the only other few recombinant vaccines currently used in humans is the Hepatitis B Virus

(HBV) vaccine, which contains a surface protein from the hepatitis virus. This protein is produced by recombinant yeast cells and then purified for injection. The HBV vaccine is much safer to use than a weakened form of the actual virus, which, if it reverts back to its original form, could cause liver cancer or hepatitis. Another recombinant vaccine that has been successfully trialled for human use is the recombinant influenza vaccine, although it has not yet been produced on a commercial basis. Feder, 2009, in fact argues that the US should have had batches of this recombinant vaccine prepared in advance during the swine flu (H1N1) outbreak in 2009, as it is fast to produce and obviates the reliance on vaccines from overseas. Recombinant influenza vaccines are composed of hemagglutinin, a protein present in various strains of the influenza virus. This protein is expressed by recombinant cell cultures and later purified to produce the vaccine.

Recombinant-Vector Vaccines

It is possible to introduce genes that encode major antigens of especially virulent pathogens into attenuated viruses or bacteria. The attenuated organism serves as a vector, replacing within the host and expressing the gene product of the pathogen. A number of organisms have been used for vector vaccines, including vaccinia virus, the canary poxvirus, attenuated poliovirus, adenoviruses, attenuated strains of *Salmonella*, and the BCG strain of *Mycobacterium bovis*. Vaccinia vector, the attenuated vaccine used to eradicate smallpox, has widely been employed as a vector vaccine. This large, complex virus, with a genome of 200 genes, can be engineered to carry several dozen foreign genes without impairing its capacity to infect host cell and replicate.

The genetically engineered vaccinia expresses high levels of the inserted gene product, which can then serve as a potent immunogen in an inoculated host. Like the smallpox vaccine, genetically engineered vaccinia vector vaccines can be administered simply by scratching the skin, causing a localized infection in host cells. If the foreign gene product expressed by the vaccinia is a viral envelope protein, it is inserted into the membrane of the infected host cell, inducing development of cell-mediated immunity as well as antibody-mediated immunity.

Other attenuated vector vaccines may prove to be safer than the vaccinia vaccine. The canary poxvirus has recently been tried as a vector vaccine. Like its relative vaccinia, the canary poxvirus is a large virus that can be easily engineered to carry multiple genes. Unlike vaccinia, the canary poxvirus does not appear to be virulent even in individuals with severe immune suppression. Another possible vector is an attenuated strain of *Salmonella typhimurium*, which has been engineered with genes from the bacterium that causes cholera. The advantage of this vector vaccine is that *Salmonella* infects cells of the mucosal lining of the gut and therefore will induce secretory IgA production. Effective immunity against a number of diseases, including cholera and gonorrhoea, depends on increased production of secretory IgA at mucous membrane surfaces. One of the poliovirus strain used in the Sabin vaccine is another candidate for a safe and effective vector vaccine. In this case, the poliovirus vector is engineered so that a portion of the gene that encodes the outer capsid protein of poliovirus is replaced by the DNA that encodes the epitope of choice. The resulting poliovirus chimera will express the desired epitope in a highly accessible presentation that protrudes from the nucleocapsid.

Recombinant Vaccines Against Bacterial Diseases

The first work of rDNA bacterial protein vaccines involved the cloning of somatic pili of enterotoxigenic E. coli (ETEC) for use in preventing diarrhoea in livestock. The ETEC have somatic pili composed of 14 to 22-KDa protein molecule that adhere to the mucosal surface of the small intestines, initiating colonization by the bacteria. Pili isolated from ETEC have been used as vaccines to prevent diarrhoeal diseases in both humans and animals. An alternate method for protecting domestic animals against ETEC has been achieved through genetic engineering. Monoclonal antibody to K99 pilus antigen produced in hybridomas and amplified in mice significantly reduced the severity of scours and the mortality of calves so treated (Sherman et al., 1983). This product was successfully tested in Canada and had since been approved by the USDA for use in the US.

One of the important impediments in the case of live vaccines is to ensure that the organism is attenuated sufficiently not to cause the disease, but still replicate to a sufficient level to induce an appropriate immune response. However, only a limited number of viral disease can be prevented by live attenuated viral vaccines state and most DNA-containing viruses have the potential to establish persistent (or latent) infection. Improvements in conventional biochemistry, recombinant DNA technology, peptide synthesis, molecular genetics and protein purification had laid the foundation for the development of new vaccines which should be more efficacious, cost effective and which have fewer side effects. Recently, recombinant DNA technology has helped to develop new generation vaccines, which are cheaper, safer and more effective. Some vaccines are made not by disarming the pathogen, but by transforming the genes coding for the antigens of pathogen into those coding for harmless characteristics. This method has been used to produce rabies vaccine. The gene Jor the surface glycoprotein of the rabies virus is inserted into the DNA of another virus vaccinia. Vaccinia virus causes cowpox, but is relatively harmless to dogs. It acts as a vector, transporting the piece of rabies virus RNA into a vaccinated individual and subsequently producing rabies antibodies. This stimulates a protective immune response against rabies.

12

DNA Chip Technology and Microarray

Microarrays are microscope slides that contain an ordered series of samples (DNA, RNA, protein, tissue). The type of microarray depends upon the material placed onto the slide: DNA (DNA microarray), RNA(RNA microarray), protein (protein microarray), tissue (tissue microarray). Since the samples are arranged in an ordered fashion, data obtained from the microarray can be traced back to any of the samples. This means that genes on the microarray are addressable. The number of ordered samples on a microarray can number into the hundred of thousands. The typical microarray contains several thousands of addressable genes. Microarray technologies as a whole provide new tools that transform the way scientific experiments are carried out. The principle advantage of microarray technologies compared with traditional methods is one of scale. In place of conducting experiments based on results from one or a few genes, microarrays allow for the simultaneous interrogation of hundreds or thousands of genes.

The most commonly used microarray is the DNA microarray. The DNA printed or spotted onto the slides can be chemically synthesized long oligonucleotides or enzymatically generated PCR products. The slides contain chemically reactive groups (typically aldehydes or primary amines) that help to stabilize the DNA onto the slide, either by covalent bonds or electrostatic interactions. An alternative technology allows the DNA to be synthesized directly onto the slide itself by a photolithographic process. This process has been commercialized and is widely available. DNA microarrays are used to determine:

1) The expression levels of genes in a sample, commonly termed expression profiling.
2) The sequence of genes in a sample, commonly termed minisequencing for short nucleotide reads, and mutation or SNP analysis for single nucleotide reads.

A DNA microarray (also commonly known as gene chip, DNA chip, or biochip) is a collection of microscopic DNA spots attached to a solid surface. Scientists use DNA microarrays to measure the expression levels of large numbers of genes simultaneously or to genotype multiple regions of a genome. A DNA microarray is a multiplex technology used in molecular biology and in medicine. It consists of an arrayed series of thousands of microscopic spots of DNA oligonucleotides, called features, each containing picomoles (10^{-12} moles) of a specific DNA sequence, known as *probes* (or *reporters*). This can be a short section of a gene or other DNA element that are used to hybridize a cDNA or cRNA sample (called *target*) under high-stringency conditions. Probe-target hybridization is usually detected and quantified by detection of fluorophore-, silver-, or chemiluminescence-labeled targets to determine relative abundance of nucleic acid sequences in the target. Since an array can contain tens of thousands of probes, a microarray experiment can accomplish many genetic tests in parallel. Therefore arrays have dramatically accelerated many types of investigation.

DNA microarrays can be used to measure changes in expression levels, to detect single nucleotide polymorphisms (SNPs), to genotype or resequence mutant genomes. Microarrays also differ in fabrication, workings, accuracy, efficiency, and cost. Additional factors for microarray experiments are the experimental design and the methods of analyzing the data (Fig. 12.1).

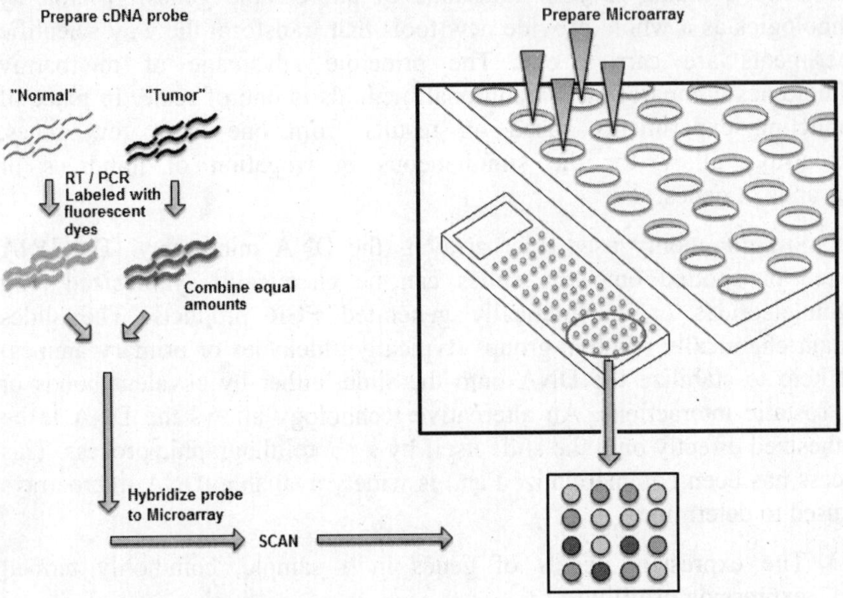

Fig. 12.1: DNA Microarray

CONSTRUCTION OF MICROARRAYS

Microarrays are unlike the traditional hybridizations assays in the substrate they use. Previous techniques used flexible membranes, for example nylon and nitrocellulose, with radioactivity and autoradiography. Microarray technology uses solid substrates like glass, with fluorescent labelling and detection. The solid substrates used in microarray technology have a number of advantages for this technique, because they are non-porous. These include:

- Deposition of minute amounts of material is possible in a precisely defined location. This is not possible with porous substrates, because of diffusion of the materials.
- Prevention of the absorption of material into the substrate, because the substrate is non-porous.
- A uniform attachment surface is provided by a solid substrate, and this improves the quality of the array elements

Comparison of the Membranes and the Microarrays assays

Criterion	Membranes	Microarrays
Surface	Porous	Non-Porous
Substrate	Non-uniform	Uniform
Conformation	Flexible	Solid
Format	Semi-Parallel	Parallel
Compatible with:		
Miniaturization	No	Yes
Photolithography	No	Yes
Piezoelectric	No	Yes
Microspotting	No	Yes
Automated production	No	Yes
Sample multiplexing	No	Yes
Sample concentration	Low	High
Hybridization Kinetics	Slow	Fast
Reagent volumes	Large	Small
Data acquisition	Slow	Fast

When compared with dot blots and Southern blots, microarrays represent high-density miniaturized arrays of molecular samples, facilitating the screening of genomic DNA or cDNA samples for the presence of one in 1, 00,000 or more DNA sequences. The technology involves hybridization of an unknown sample to an ordered array of immobilized molecules of known sequences. This produces a specific hybridization pattern that can be analyzed or compared to a given standard sometimes with the help of computer devices. These microarrays can be prepared either by the synthesis of oligonucleotides on a DNA chip or by deposition of available DNA sequences (e.g. PCR products, cloned DNA fragments) on such a chip. Miniaturization of conventional assays (as in Southern blots for RFLP) is a general trend in molecular biology, since micro scale assays reduce reagent consumption, minimize reaction volumes, increase the sample concentration and accelerate the reaction kinetics. Although the initial microarray assays focused on nucleic acid hybridization, future studies will certainly involve parallel analysis of proteins, lipids, carbohydrates and other small molecules. For instance, protein chips have already become available for the study of proteins.

Types of DNA Chips and Their Production

Following two types of DNA chips are generally available:

(i) **Oligonucleotide-based chips,** which contain high density of short oligonucleotides of known sequences, generally 20 to 25 nucleotides long (100,000 to 400,000 oligonucleotides immobilized within an area of 1.6 cm^2).

(ii) **cDNA-based chips**, which contain a high density of cDNA samples, sometimes derived as PCR amplified DNA fragments. While the former are often used for **sequencing by hybridization (SBH)** for detection of SNPs, etc., the latter are often used for studying gene expression patterns in time and space (for SNPs, consult).

Production of oligonucleotide microarrays

In oligonucleotide microarrays, the probes are short sequences designed to match parts of the sequence of known or predicted open reading frames. Although oligonucleotide probes are often used in "spotted" microarrays, the term "oligonucleotide array" most often refers to a specific technique. Oligonucleotide arrays are produced by printing short oligonucleotide sequences designed to represent a single gene or family of gene splice-variants by synthesizing this sequence directly onto the array surface instead of depositing intact sequences. Sequences may be longer (60-mer probes) or shorter (25-mer probes) depending on the desired purpose; longer probes are more specific to individual target genes, and shorter probes may be spotted across the array in higher densities and are cheaper to manufacture.

One technique used to produce oligonucleotide arrays involves photolithographic synthesis on a silica substrate where light and light-sensitive masking agents are used to build a sequence one nucleotide at a time across the entire array. Each applicable probe is selectively unmasked prior to bathing the array in a solution containing just one nucleotide. Then a masking reaction takes place and the next set of probes is unmasked in preparation for a different nucleotide exposure. After many repetitions, the sequence of every probe becomes complete. More recently, Maskless Array Synthesis from NimbleGen Systems has adapted the flexibility of this system with other methods to produce larger numbers of probes.

Oligonucleotide microarrays, can be produced through DNA synthesis on solid surfaces (glass surface is of then used) using combinatorial chemistry methods (for combinatorial chemistry, consult). Since in such an in situ method, it is not possible to purify the full length reaction product, this in situ synthesis puts a limitation on the length of oligonucleotides that can be synthesized. However, advantage of the technique is, that large number of oligonucleotides, at many sites, can be synthesized simultaneously, and data can be collected without physically handling the clones. This also allows a very high density of microarrays. Two different techniques are known for the production of oligonucleotide microarrays.

Light directed deprotection method

The production of oligonucleotide microarrays makes use of a combination of photolithography with DNA-synthetic chemistry. In this technique, nucleotides are used in the form of modified phosphoramidites (as used in DNA synthesis machines), having a photo-labile protecting group, so that these protecting groups can be removed by exposing nucleotides to light. In other words, light is used as an activating agent in the synthesis reaction (Fig. 12.2). Photolithographic masks are used to control the regions of the chip that are illuminated but not desired to be deprotected. After deprotection the surface of the chip is flooded with one of the four bases to allow selective coupling of that base to each deprotected region of the synthesis surface. A seconded region of the chip is then deprotected and similarly flooded with another base for coupling reaction. In this manner, four cycles will be needed for adding the first base of each of the thousands of oligonucleotides being synthesized. Since for addition of one base to each oligomer, four cycles are needed, a maximum of 20 x 4 = 80 cycles will be needed for the production of all possible 20 mers (4^{20}). This method allows the production of very high density (250000 features/cm^2) microarrays. However, the synthesis of oligonucleotides longer than 25 mers is difficult with the currently available efficiency of reactions. In future, improvement in photolithographic technology may allow manufacture of microarrays with oligonucleotides containing >25 bases without the need of photomasks.

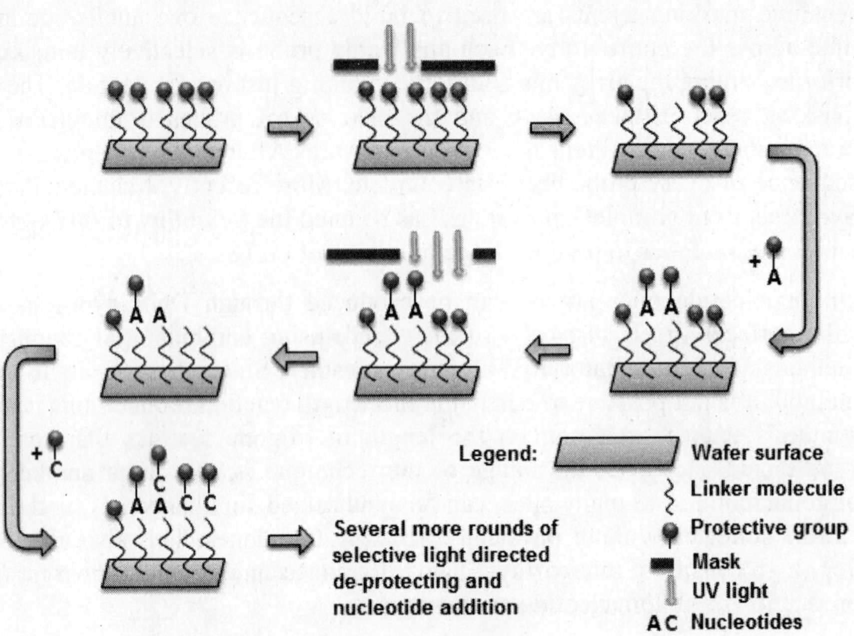

Fig. 12.2: Light directed deprotection method

Ink-jet printing technology

Oligonucleotide micro arrays can also be produced by ink-jet printing technology, which involves direct delivery of phosphoramidites, rather than the use of photolithographic masks (Fig. 12.3). Following steps are involved:

(i) Glass surface is coated with a light sensitive hydrophobic material.

(ii) Synthetic areas are prepared by photolithographic etching.

(iii) Phosphoramidites are delivered at specific hydrophilic wells created as above, and deblocking agent is used to oxidize and complete one round of coupling.

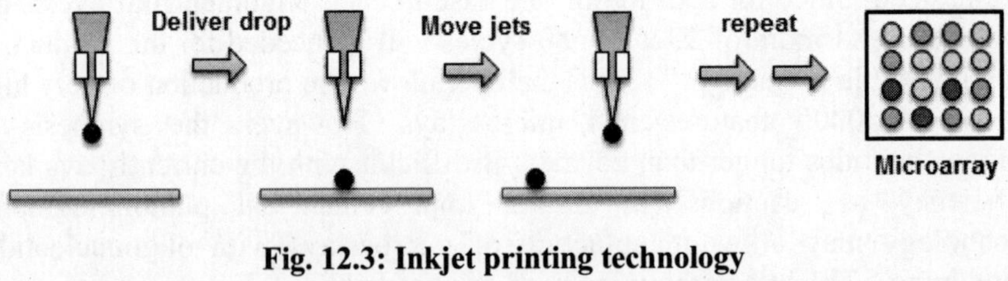

Fig. 12.3: Inkjet printing technology

The cycle is repeated by delivering specific phosphoramidites at specific addresses. This ink-jetting is simpler, so that it may allow oligonucleotide micro arrays to reach a large number of researchers. However, only modest densities (10,000 features/cm^2) are attainable by this technology. Microarrays using this technology are being commercialized by Protogene, Incyte, Hewlett Packard, Rosetta and several other companies.

Spotted microarrays

The probes are oligonucleotides, cDNA, or small fragments of PCR products that correspond to mRNAs. Each probe contains a different, characteristic sequence that is specific to a different group of genes under study. These probes are then spotted onto glass substrate to form an array. One common approach utilizes an array of fine pins or needles controlled by a robotic arm that is dipped into wells containing different DNA probes (Fig. 12.4). Each needle then deposits its probe onto designated locations on the array surface. The probes are then ready to hybridize with complementary cDNA and cRNA targets derived from experimental or clinical samples. This technique is used by research scientists around the world to produce "in-house" printed microarrays from their own labs. These arrays can be customized to each experiment easily because the researchers can choose any set of probes and printing locations, synthesize the probes in their own labs (or collaborating facilities), and spot the arrays themselves. They can then generate their own labeled samples for hybridization, hybridize the samples to the array, and finally scan the arrays with their own equipment. This provides a relatively low-cost microarray well customized to each study, avoiding the cost of more expensive commercial arrays that often represent vast numbers of genes that are not of interest to the investigator.

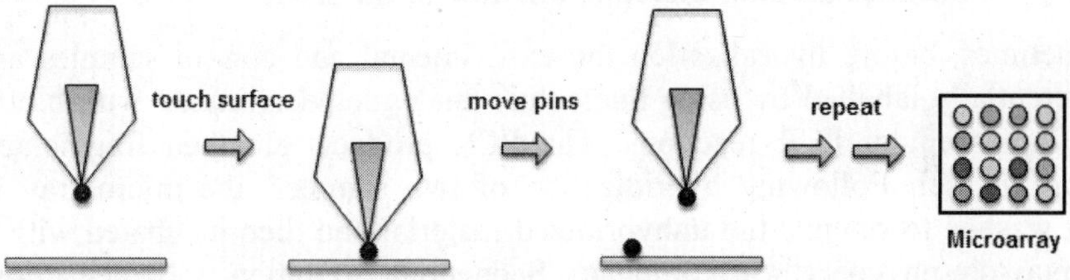

Fig. 12.4: Spotting technology

Production of cDNA microarrays

Production of microarrays involving large cDNA fragments (molecules, much larger than 20-mer oligonucleotides) involves delivery by depositing systems rather than *in situ* synthesis. Small amounts of specific cDNA fragments are deposited onto known locations on glass surfaces. Two liquid delivery technologies have been used extensively for the production of cDNA microarrays.

Hybridization onto DNA Chips

Once the DNA chips with microarrays are produced, the next step is hybridization of an unknown sample (under investigation) to these DNA chips. Mixtures of DNA or RNA isolated from biological systems are labelled enzymatically by incorporating nucleotides bearing reporter tags and hybridized to microarrays. Hybridization reactions yield heteroduplexes between individual components of the fluorescent probe and the complementary target sequence on the chip. Since each target sequence is chemically homogenous and occupies a known position, the identity and quantity of each component in the unknown sample can be ascertained by measuring the intensity of the fluorescence at each position on the microarray. Sophisticated fluorescence technology is used for detection, which includes confocal laser scanning and charge coupled device (CCD) imaging, which allow collection of quantitative data at a fast speed. Direct detection of probe binding by electronic means is also possible. However, hybridization is affected by the following factors:

(i) Influence of the dangling ends of a bound molecule on the stability of the heteroduplex.

(ii) The degree of accessibility of the probe to hybridization due to intrastrand secondary and tertiary structures: this can be partly resolved by fragmentation of the probe, although it may interfere with labeling.

(iii) Concentration of probe also effects hybridization; electric fields are sometimes used for directing the flow of the probe.

Sometimes, before hybridization the experimental and control samples are differentially labelled by using nucleotide analogues derivatized with biotin ad fluorescein in PCR reactions. The PCR products are then fragmented using DNaseI. Following hybridization of two samples, the microarray is first washed to remove the unhybridized material and then incubated with a streptavidin-phycoerythrin conjugate. Sequential excitation will then allow the detection of phycoerythrin and fluorescein emissions.

Types of Microarrays

Depending upon the kind of immobilized sample used construct arrays and the information fetched, the Microarray experiments can be categorized in three ways:

Microarray expression analysis

In this experimental setup, the cDNA derived from the mRNA of known genes is immobilized. The sample has genes from both the normal as well as the diseased tissues. Spots with more intensity are obtained for diseased tissue gene if the gene is over expressed in the diseased condition. This expression pattern is then compared to the expression pattern of a gene responsible for a disease.

Microarray for mutation analysis

For this analysis, the researchers use gene DNA. The genes might differ from each other by as less as a single nucleotide base. A single base difference between two sequences is known as Single Nucleotide Polymorphism (SNP) and detecting them is known as SNP detection.

Comparative Genomic Hybridization

It is used for the identification in the increase or decrease of the important chromosomal fragments harboring genes involved in a disease.

Different types of microarray-based assays:

- DNA microarrays for transcriptomics research
- DNA microarrays for genotyping
- DNA microarrays for resequencing
- DNA tiling arrays for ChIP on chip experiments
- Protein microarrays to detect protein-protein interactions
- Antibody-Based Protein Array for protein quantification
- Chromatin Immunoprecipitation Assay (ChIP)
- Carbohydrate microarrays
- Tissue Microarray
- Living Cell Arrays

Applications of Microarrays

Gene discovery

DNA Microarray technology helps in the identification of new genes, know about their functioning and expression levels under different conditions.

Disease diagnosis

DNA Microarray technology helps researchers learn more about different diseases such as heart diseases, mental illness, infectious disease and especially the study of cancer. Until recently, different types of cancer have been classified on the basis of the organs in which the tumors develop. Now, with the evolution of microarray technology, it will be possible for the researchers to further classify the types of cancer on the basis of the patterns of gene activity in the tumor cells. This will tremendously help the pharmaceutical community to develop more effective drugs as the treatment strategies will be targeted directly to the specific type of cancer.

Drug discovery

Microarray technology has extensive application in Pharmacogenomics. Pharmacogenomics is the study of correlations between therapeutic responses to drugs and the genetic profiles of the patients. Comparative analysis of the genes from a diseased and a normal cell will help the identification of the biochemical constitution of the proteins synthesized by the diseased genes. The researchers can use this information to synthesize drugs which combat with these proteins and reduce their effect.

Toxicological research

Microarray technology provides a robust platform for the research of the impact of toxins on the cells and their passing on to the progeny. Toxicogenomics establishes correlation between responses to toxicants and the changes in the genetic profiles of the cells exposed to such toxicants.

DNA microarrays and proteomics

Like genomics, the term 'proteomics' relates to the study of proteins and the protein-protein interactions. DNA chips can also be used for this purpose. Protein linkage map can also be created using genomic sequence information. Protein-protein interactions are generally studied by yeast two hybrid system. In this system genomic clones of two fusion proteins are used for the activation of transcription of a reporter gene in yeast.

The first fusion protein contains a DNA binding domain fused to a protein of interest and the second fusion protein carries an acidic transcription activation domain fused to the protein, whose interaction with the first protein of interest is being examined. Scientific interaction between two chimeric (fused) proteins leads to transcriptional activation, thus confirming protein-protein interaction. The identity of the two proteins of interest is confirmed by sequence analysis of each clone thus identified. Therefore, major sequencing work is involved in the above two hybrid system.

Determining tissue-specific gene expression patters

The most straightforward application of microarrays is to characterize the transcriptome of a particular cell type or tissue. The patterns of gene expression obtained are specific for the differentiation state of the cells or tissues under study. In this context microarray analysis could be used as a phenotyping tool for embryonic cells to determine whether a differentiation protocol has been successful. Microarray analysis would provide a comprehensive molecular phenotype of the differentiation state of the cells.

Determining response to treatments

Microarray analysis enables the global assessment of drug effects in appropriate cell, or animal model systems or patient samples. Besides steady-state effects, it is also possible to determine the kinetics of the expression response as different genes respond in a time-dependent fashion at different intervals following a perturbation.

Determining specific effects of compounds

Microarray analysis can be used to determine or differentiate between different types of drug specific effects on model systems or patient samples. The results may lead to the identification of new drug targets through the study of disease models.

Determining toxic effects of compounds

Microarray analysis makes it possible to monitor in a comprehensive way the toxic effects that compounds may have on cells, tissue or patient samples. Determining the effects of model compound may enable prediction as to putative toxic effects of novel compound early on in the drug development pipeline.

Stratification of patient groups

Microarray analysis will facilitate the identification of groups of patients that respond similarly to a particular treatment regime. Determining comprehensively how cells or tissues respond to a particular drug might make it possible to predict either efficacy or adverse effects to a drug.

Provide insights into gene function

For many genes, we still do not have any indication as to what their function could be. As genes that have related functions often share expression patterns, it is reasonable to assume that microarray data can be used to investigate functional relationships between genes. For many novel genes discovered as a result of the genome projects, microarray results represent the fist clue to the question of what these genes might do. Widespread expression suggests a ubiquitous or housekeeping function, whereas cell-, tissue- or organ-specific expression implies a function related to that of the specific cell, tissue or organ in which that gene is specifically expressed. As a matter of fact, even an undetectable expression level in a particular cell type can be a very useful piece of information as it excludes a role for that gene in a cell type where it is not expressed in.

Determination of molecular pathways

Microarray analysis of gene-expression changes in biological systems that model disease might lead to the identification of pathways that play a crucial role in disease pathogenesis.

Determine common regulatory mechanisms

Microarray analysis can be used to identify groups of genes that show similar response to particular physiological conditions or perturbations. Analysis of promoted regions of members of these groups of genes are likely to reveal common cis-acting DNA sequences that are responsible for the coordinated response of the genes.

Determine metastatic potential of cancer cells

Microarray analysis will provide insight into the pattern of gene expression that allows tumors to metastasize, by comparing gene expression profiles of tumor variants with low or high metastatic potential.

Identification of biomarkers

Microarray analysis provides the ability to identify biomarkers that can be used to readily monitor drug efficacies as well as side effects and disease progression.

Problems associated with high throughput gene expression analysis

The strongest quality of DNA microarrays is the enormous amount of genes that can be studied at once. Paradoxically, the source of its power is also one of the weak points of the technology as we still do not have the right statistical know-how and tools to deal with the multiple testing problems associated with performing 50000 tests at the same time. This is also referred to as "the curse of multidimensionality of microarray data." Another problem is the remarkable feature of microarray data sets.

Most conventional data sets for which we have over a hundred years of experience in dealing with them are data sets in which scientists have looked at one or two variables and up to hundreds or even many thousands of samples. Just think of a statistician trying to predict who will win the next presidential elections (just two candidates) by interviewing many thousands of voters, or an analysis of the effect of soil type on the crop yield based on counting thousands of maize grains.

13

Genome Mapping

GENOME

A genome is all of a living thing's genetic material. It is the entire set of hereditary instructions for building, running, and maintaining an organism, and passing life on to the next generation. In most living things, the genome is made of a chemical called DNA. The genome contains genes, which are packaged in chromosomes and affect specific characteristics of the organism. In short, the genome is divided into chromosomes, chromosomes contain genes, and genes are made of DNA. The word "genome" was coined in about 1930, even though scientists didn't know then what the genome was made of. They only knew that the genome was important enough, whatever it was, to have a name. Each one of earth's species has its own distinctive genome: the dog genome, the wheat genome, the genomes of the cow, cold virus, *Escherichia coli* (a bacterium that lives in the human gut and in animal intestines), and so on. So genomes belong to species, but they also belong to individuals. Every giraffe on the African savanna has a unique genome, as does every elephant, acacia tree, and ostrich. Unless there is identical twin, genome is different from that of every other person on earth—in fact, it is different from that of every other person who has ever lived.

Genomes are found in cells, the microscopic structures that make up all organisms. With a few exceptions, each of our body's trillions of cells contains a copy of our genome: the cells in our muscles, the cells in our brain, the cells in our blood, and so on. Imagine all the trillions of genomes in our body, in other people's bodies, in cedar and apple trees, in walruses, forest mushrooms, and migrating birds: The whole world is full of genomes.

GENE

A gene is a small piece of the genome. It's the genetic equivalent of the atom: As an atom is the fundamental unit of matter, a gene is the fundamental unit of heredity. Genes are found on chromosomes and are made of DNA. Different genes determine the different characteristics, or traits, of an organism. In the simplest terms (which are actually too simple in many cases), one gene might determine the color of a bird's feathers, while another gene would determine the shape of its beak. The number of genes in the genome varies from species to species. More complex organisms tend to have more genes. Bacteria have several hundred to several thousand genes. All individuals in a species have the same set of genes: in peas there is a gene for pod color, a gene for plant height, a gene for pea shape, and so on. What makes individuals different is that a gene can have several different forms, or alleles. Thus, in peas, the pod color gene has green and yellow alleles, the plant height gene has tall and dwarf alleles, and so on. Individuals have two copies of each gene, one inherited from each parent. How the two copies interact with each other determines an organism's characteristics. It doesn't matter whether a gene comes from the maternal or paternal side—a green-pod allele from a sperm cell is the same as a green-pod allele from an egg cell. What matters is how many of which type of allele an individual gets. Genes tell a cell how to make proteins. Roughly speaking, each gene is a set of instructions for making one specific protein.

Proteins are a diverse group of large, complex molecules that are crucial to every aspect of the body's structure and function. Collagen, which forms the structural scaffolding of skin and many other tissues, is a protein. Insulin, a hormone that regulates blood sugar, is a protein. Trypsin, an enzyme involved in digestion, is a protein. So is the pigment melanin, which gives hair and skin its color. Still other proteins regulate the body's production of proteins. Genes sometimes affect characteristics in indirect ways. For example, genes affect the size and shape of our nose, even though there's no such thing as a "nose size" protein. But directly or indirectly, the way genes influence our traits is by telling our cells which proteins to make, how much, when, and where.

A gene has several parts. In most genes, the protein-making instructions are broken up into relatively short sections called exons. These are interspersed with introns, longer sections of "extra" or "nonsense" DNA. Genes also contain regulatory sequences, which help determine where, when, and in what amount proteins are made. Though still poorly understood, these regulatory sequences are crucial to how our body works. They help determine which genes are "turned on," or transmitting their protein-making instructions to the rest of the cell, in different cells throughout the body. All cells contain the same genes, but cells don't make all the proteins they have genes for. Thus, what makes a liver cell different from a brain cell is which proteins it makes—and therefore, which of its genes are turned on.

GENOME MAP

A genome map helps scientists navigate around the genome. Like road maps and other familiar maps, a genome map is a set of landmarks that tells people where they are, and helps them get where they want to go. The landmarks on a genome map might include short DNA sequences, regulatory sites that turn genes on and off, and genes themselves. Often, genome maps are used to help scientists find new genes. Some parts of the genome have been mapped in great detail, while others remain relatively uncharted territory. It may turn out that a few landmarks on current genome maps appear in the wrong place or at the wrong distance from other landmarks. But over time, as scientists continue to explore the genome frontier, maps will become more accurate and more detailed. A genome map is a work in progress.

Most everyday maps have length and width, latitude and longitude, like the world around us. But a genome map is one-dimensional—it is linear, like the DNA molecules that make up the genome itself. A genome map looks like a straight line with landmarks noted at irregular intervals along it, much like the towns along the map of a highway. The landmarks are usually inscrutable combinations of letters and numbers that stand for genes or other features—for example, D14S72, GATA-P7042, and so on.

Difference between a genome map and a genome sequence

Both are portraits of a genome, but a genome map is less detailed than a genome sequence. A sequence spells out the order of every DNA base in the genome, while a map simply identifies a series of landmarks in the genome. Sometimes mapping and sequencing are completely separate processes. For example, it's possible to determine the location of a gene—to "map" the gene—without sequencing it. Thus, a map may tell you nothing about the sequence of the genome, and a sequence may tell you nothing about the map.

In other cases, the landmarks on a map are DNA sequences, and mapping is the cousin of sequencing. For example, consider the following sequence:

GCCATTGACGTCCCCTTGAAA

CGGTAACTGCAGGGGAACTTT

A map of that sequence might look like this:

GCC--------CCCC------

On this map, GCC is one landmark; CCCC is another. In the corresponding sequence, each base is a landmark. In other words, the sequence is simply the most detailed possible map. In general, particularly for humans and other species with large genomes, creating a reasonably comprehensive genome map is quicker and cheaper than sequencing the entire genome. Simply put, mapping involves less information to collect and organize than sequencing does. Many animal "genome projects" currently underway, such as those that focus on the dog and the horse, aim to map the genomes of these species. This will help scientists learn more about the biology of these species, without the enormous resources required when sequencing a genome. By contrast, studying the human genome is actually a two-pronged effort, aiming at both a comprehensive genome map and a complete genome sequence. Advances in sequencing help the mappers move ahead, and advances in mapping help the sequencers make progress. These efforts are closely linked but not exactly the same thing.

At first glance, this strategy of sequencing and mapping may seem redundant, since a sequence is simply the most detailed map possible. Why not just sequence the genome? And why keep mapping the human genome if it's already been sequenced? One reason is that a map can actually help us sequence the genome. If we're sequencing a genome with the clone-by-clone method, we need a map in order to determine where each clone belongs in the genome. The more detailed and accurate our map, the easier it is to snap those pieces of genomic jigsaw puzzle into place. With whole-genome shotgun sequencing, a map is no longer central to the strategy, but one can still be used to help match the big pieces of assembled sequence to their proper place in the genome.

Another reason—which also helps explain why the human genome sequence doesn't make the mappers' work moot—is that you need a map in order to understand the genome sequence. A sequence is pretty much featureless: just a long, long string of DNA bases or "letters." For the most part, scientists can't look at a sequence and see immediately which parts are genes or other interesting features, and which parts are "junk." But the landmarks on a genome map provide clues about where the important parts of the genome sequence can be found.

Uses of Genome Maps

Genome maps help scientists find genes, particularly those involved in human disease. This process is much like a scientific game of hot and cold. Scientists study many families affected by a disease, tracing the inheritance of the disease and of specific genome landmarks through several generations. Landmarks that tend to be inherited along with the disease are likely to be located close to the disease gene and become "markers" for the gene in question. Once they have identified a few such markers, scientists know the approximate location of the disease gene. In this way, they narrow down their search from the entire 3-billion-base-pair genome to a region of the genome a few million base pairs long. Next, they look for genes in that part of the genome and study the genes one by one to learn which one is involved in the disease. For example, they might look for a gene that has a different sequence in people with the disease than it does in healthy people. Or they might look for a gene with a function that could be related to the disease. Genes for cystic fibrosis, Huntington's disease, and many other inherited diseases have been identified by this method. But it's a time-consuming, laborious process. Several million base pairs is still a lot of DNA, and a region of the genome that size may contain dozens of genes for scientists to sort through.

In addition to helping in the search for genes, genome maps are useful in the day-to-day activities of molecular biology laboratories. In the lab, the human genome lives in the form of "clones"—chunks of DNA that have been chopped up and spliced into the DNA of bacteria or other cells. This method keeps each chunk of the genome separate from the others and available in many copies for easy experiment and study. Genome maps also help scientists find and learn about other important parts of the genome, such as the regulatory regions that help control when genes are turned on and off. Maps help scientists keep track of which colleagues are studying nearby or related parts of the genome, so they can learn from each other and don't duplicate each other's work. They illuminate the overall structure of the genome—places where several related genes are clustered together, for example, or parts of the genome that contain an unusually rich concentration of genes. Finally, genome maps enable scientists to compare the genomes of different species, yielding insights into the process of evolution.

Types of genome Maps

The two main kinds of genome maps are known as genetic-linkage maps and physical maps. Both genetic and physical maps provide the likely order of items along a chromosome. However, a genetic map, like an interstate highway map, provides an indirect estimate of the distance between two items and is limited to ordering certain items. One could say that genetic maps serve to guide a scientist toward a gene, just like an interstate map guides a driver from city to city. On the other hand, physical maps mark an estimate of the true distance, in measurements called base pairs, between items of interest. To continue our analogy, physical maps would then be similar to street maps, where the distance between two sites of interest may be defined more precisely in terms of city blocks or street addresses. Physical maps, therefore, allow a scientist to more easily home in on the location of a gene. An appreciation of how each of these maps is constructed may be helpful in understanding how scientists use these maps to traverse that genetic highway commonly referred to as the "human genome".

Genetic Maps

Genetic maps serve to guide a scientist toward a gene, just like an interstate map guides a driver from city to city. Genetic maps have landmarks known as genetic markers, or "markers" for short. The term marker is used very broadly to describe any observable variation that results from an alteration, or mutation, at a single genetic locus. A marker may be used as one landmark on a map if, in most cases, that stretch of DNA is inherited from parent to child according to the standard rules of inheritance. Markers can be within genes that code for a noticeable physical characteristic such as eye color, or a not so noticeable trait such as a disease. DNA-based reagents can also serve as markers. These types of markers are found within the non-coding regions of genes and are used to detect unique regions on a chromosome. DNA markers are especially useful for generating genetic maps when there are occasional, predictable mutations that occur during meiosis—the formation of gametes such as egg and sperm—that, over many generations, lead to a high degree of variability in the DNA content of the marker from individual to individual.

Genetic-linkage mapping

Genetic-linkage maps illustrate the order of genes on a chromosome and the relative distances between those genes. Originally, these maps were made by tracing the inheritance of multiple traits, such as hair color and eye color, through several generations. Genetic-linkage mapping is possible because of a normal biological process called crossing over, which occurs during meiosis—a type of cell division for making sperm and egg cells. During one stage of meiosis, chromosomes line up in pairs along the center of a cell, where they sometimes "stick" to each other and exchange equivalent pieces of themselves. This sticking and exchanging is called crossing over, and is a relatively common event: on average, a chromosome pair undergoes crossing over about 1.5 times during the formation of each sex cell in humans.

For example, imagine a man who has one chromosome with brown-eye and brown-hair genes and another chromosome with blue-eye and blonde-hair genes in his cells. Usually, his sperm cells will have either brown-eye and brown-hair genes, or blue-eye and blonde-hair genes. But if crossing over occurs, the man will produce one sperm cell with brown-eye and blonde-hair genes, and another with blue-eye and brown-hair genes. In short, crossing over produces chromosomes with new combinations of genes—and offspring, called recombinants, with new combinations of traits not seen in either parent. Generally the closer two genes are on a chromosome; the less likely they are to be separated by crossing over.

This means that traits that are inherited together most often are probably influenced by genes that are close to each other on a chromosome. On the other hand, traits that are inherited together less often are probably influenced by genes that are farther apart. Thus, by following several traits through generations and recording how often recombinants occur, one can map the relative position of corresponding genes. Early geneticists recognized that genes are located on chromosomes and believed that each individual chromosome was inherited as an intact unit. They hypothesized that if two genes were located on the same chromosome; they were physically linked together and were inherited together. We now know that this is not always the case. Studies conducted around 1910 demonstrated that very few pairs of genes displayed complete linkage. Pairs of genes were either inherited independently or displayed partial linkage—that is, they were inherited together sometimes, but not always.

During meiosis—the process whereby gametes (eggs and sperm) are produced— two copies of each chromosome pair become physically close. The chromosome arms can then undergo breakage and exchange segments of DNA, a process referred to as recombination or crossing-over. If recombination occurs, each chromosome found in the gamete will consist of a "mixture" of material from both members of the chromosome pair. Thus, recombination events directly affect the inheritance pattern of those genes involved. Because one cannot physically see crossover events, it is difficult to determine with any degree of certainty how many crossovers have actually occurred. But, using the phenomenon of co-segregation of alleles of nearby markers, researchers can reverse-engineer meiosis and identify markers that lie close to each other. Then, using a statistical technique called genetic linkage analysis, researchers can infer a likely crossover pattern, and from that an order of the markers involved. Researchers can also infer an estimate for the probability that a recombination occurs between each pair of markers.

If recombination occurs as a random event, then two markers that are close together should be separated less frequently than two markers that are more distant from one another. The recombination probability between two markers, which can range from 0 to 0.5, increases monotonically as the distance between the two markers increases along a chromosome. Therefore, the recombination probability may be used as a surrogate for ordering genetic markers along a chromosome. If you then determine the recombination frequencies for different pairs of markers, you can construct a map of their relative positions on the chromosome. Predicting recombination is not so simple. Although crossovers are random, they are not uniformly distributed across the genome or any chromosome. Some chromosomal regions, called recombination hotspots, are more likely to be involved in crossovers than other regions of a chromosome. This means that genetic map distance does not always indicate physical distance between markers. Despite these qualifications, linkage analysis usually correctly deduces marker order, and distance estimates are sufficient to generate genetic maps that can serve as a valuable framework for genome sequencing.

Linkage Studies in Patient Populations: Genetic Maps and Gene Hunting

In humans, data for calculating recombination frequencies are obtained by examining the genetic makeup of the members of successive generations of existing families, termed human pedigree analysis. Linkage studies begin by obtaining blood samples from a group of related individuals. For relatively rare diseases, scientists find a few large families that have many cases of the disease and obtain samples from as many family members as possible. For more common diseases where the pattern of disease inheritance is unclear, scientists will identify a large number of affected families and will take samples from four to thirty close relatives. DNA is then harvested from all of the blood samples and screened for the presence, or co-inheritance, of two markers. One marker is usually the gene of interest, generally associated with a physically identifiable characteristic. The other is usually one of the various detectable rearrangements mentioned earlier, such as a microsatellite. A computerized analysis is then performed to determine whether the two markers are linked and approximately how far apart those markers are from one another. In this case, the value of the genetic map is that an inherited disease can be located on the map by following the inheritance of a DNA marker present in affected individuals but absent in unaffected individuals, although the molecular basis of the disease may not yet be understood, nor the gene(s) responsible identified.

Commonly Used DNA markers:

- **RFLPs**, or **restriction fragment length polymorphisms**, were among the first developed DNA markers. RFLPs are defined by the presence or absence of a specific site, called a restriction site, for a bacterial restriction enzyme. This enzyme breaks apart strands of DNA wherever they contain a certain nucleotide sequence.

- **VNTRs**, or **variable number of tandem repeat polymorphisms**, occur in non-coding regions of DNA. This type of marker is defined by the presence of a nucleotide sequence that is repeated several times. In each case, the number of times a sequence is repeated may vary.

- **Microsatellite polymorphisms** are defined by a variable number of repetitions of a very small number of base pairs within a sequence. Oftentimes, these repeats consist of the nucleotides, or bases, cytosine and adenosine. The number of repeats for a given microsatellite may differ between individuals, hence the term **polymorphism**--the existence of different forms within a population.

- **SNPs**, or **single nucleotide polymorphisms**, are individual point mutations, or substitutions of a single nucleotide, that do not change the overall length of the DNA sequence in that region. SNPs occur throughout an individual's genome.

Genetic Maps as a Framework for Physical Map Construction

Genetic maps are also used to generate the essential backbone, or scaffold, needed for the creation of more detailed human genome maps. These detailed maps, called **physical maps**, further define the DNA sequence between genetic markers and are essential to the rapid identification of genes.

Physical mapping

Physical maps, by contrast, always give the physical, DNA-base-pair distances from one landmark to another. In the late 1970s, scientists developed new and efficient ways of cutting the genome up into smaller pieces in order to study it. Around the same time they made the first physical maps by using the overlapping DNA sequences at the ends of the genome pieces to help them keep track of where the pieces came from. (The process had a lot in common with the assembly step of genome sequencing.) In other words, a physical map was simply an ordered set of DNA pieces.

Types of Physical Maps and What They Measure

Physical maps can be divided into three general types: chromosomal or cytogenetic maps, radiation hybrid (RH) maps, and sequence maps. The different types of maps vary in their degree of resolution, that is, the ability to measure the separation of elements that are close together. The higher the resolution, the better the picture. The lowest-resolution physical map is the chromosomal or cytogenetic map, which is based on the distinctive banding patterns observed by light microscopy of stained chromosomes. As with genetic linkage mapping, chromosomal mapping can be used to locate genetic markers defined by traits observable only in whole organisms. Because chromosomal maps are based on estimates of physical distance, they are considered to be physical maps. Yet, the number of base pairs within a band can only be estimated.

RH maps and sequence maps, on the other hand, are more detailed. RH maps are similar to linkage maps in that they show estimates of distance between genetic and physical markers, but that is where the similarity ends. RH maps are able to provide more precise information regarding the distance between markers than can a linkage map. The physical map that provides the most detail is the sequence map. Sequence maps show genetic markers, as well as the sequence between the markers, measured in base pairs.

RH Mapping

RH mapping, like linkage mapping, shows an estimated distance between genetic markers. But, rather than relying on natural recombination to separate two markers, scientists use breaks induced by radiation to determine the distance between two markers. In RH mapping, a scientist exposes DNA to measured doses of radiation, and in doing so, controls the average distance between breaks in a chromosome. By varying the degree of radiation exposure to the DNA, a scientist can induce breaks between two markers that are very close together. The ability to separate closely linked markers allows scientists to produce more detailed maps. RH mapping provides a way to localize almost any genetic marker, as well as other genomic fragments, to a defined map position, and RH maps are extremely useful for ordering markers in regions where highly polymorphic genetic markers are scarce.

Polymorphic refers to the existence of two or more forms of the same gene, or genetic marker, with each form being too common in a population to be merely attributable to a new mutation. Polymorphism is a useful genetic marker because it enables researchers to sometimes distinguish which allele was inherited. Scientists also use RH maps as a bridge between linkage maps and sequence maps. In doing so, they have been able to more easily identify the location(s) of genes involved in diseases such as spinal muscular atrophy and hyperekplexia, more commonly known as "startle disease".

Sequence Mapping

Sequence tagged site (STS) mapping is another physical mapping technique. An STS is a short DNA sequence that has been shown to be unique. To qualify as an STS, the exact location and order of the bases of the sequence must be known, and this sequence may occur only once in the chromosome being studied or in the genome as a whole if the DNA fragment set covers the entire genome.

Common sources of STSs

- **Expressed sequence tags (ESTs)** are short sequences obtained by analysis of complementary DNA (cDNA) clones. Complementary DNA is prepared by converting mRNA into double-stranded DNA and is thought to represent the sequences of the genes being expressed. An EST can be used as an STS if it comes from a unique gene and not from a member of a gene family in which all of the genes have the same, or similar, sequences.
- **Simple sequence length polymorphisms (SSLPs)** are arrays of repeat sequences that display length variations. SSLPs that are polymorphic and have already been mapped by linkage analysis are particularly valuable because they provide a connection between genetic and physical maps.
- **Random genomic sequences** are obtained by sequencing random pieces of cloned genomic DNA or by examining sequences already deposited in a database.

To map a set of STSs, a collection of overlapping DNA fragments from a chromosome is digested into smaller fragments using restriction enzymes, agents that cut up DNA molecules at defined target points. The data from which the map will be derived are then obtained by noting which fragments contain which STSs. To accomplish this, scientists copy the DNA fragments using a process known as "molecular cloning". Cloning involves the use of a special technology, called recombinant DNA technology, to copy DNA fragments inside a foreign host.

First, the fragments are united with a carrier, also called a vector. After introduction into a suitable host, the DNA fragments can then be reproduced along with the host cell DNA, providing unlimited material for experimental study. An unordered set of cloned DNA fragments is called a library.

Next, the clones, or copies, are assembled in the order they would be found in the original chromosome by determining which clones contain overlapping DNA fragments. This assembly of overlapping clones is called a clone contig. Once the order of the clones in a chromosome is known, the clones are placed in frozen storage, and the information about the order of the clones is stored in a computer, providing a valuable resource that may be used for further studies. These data are then used as the base material for generating a lengthy, continuous DNA sequence, and the STSs serve to anchor the sequence onto a physical map.

The Need to Integrate Physical and Genetic Maps

As with most complex techniques, STS-based mapping has its limitations. In addition to gaps in clone coverage, DNA fragments may become lost or mistakenly mapped to a wrong position. These errors may occur for a variety of reasons. A DNA fragment may break, resulting in an STS that maps to a different position. DNA fragments may also get deleted from a clone during the replication process, resulting in the absence of an STS that should be present. Sometimes a clone composed of DNA fragments from two distinct genomic regions is replicated, leading to DNA segments that are widely separated in the genome being mistakenly mapped to adjacent positions. Lastly, a DNA fragment may become contaminated with host genetic material, once again leading to an STS that will map to the wrong location. To help overcome these problems, as well as to improve overall mapping accuracy, researchers have begun comparing and integrating STS-based physical maps with genetic, RH, and cytogenetic maps. Cross-referencing different genomic maps enhances the utility of a given map, confirms STS order, and helps order and orient evolving contigs.

As mapping techniques advance, scientists try to create maps with more landmarks that are more closely, evenly, and accurately spaced. But in contrast to DNA sequencing, which has become increasingly automated, genome mapping still can only be accomplished by experienced scientists. And even the most expert mapper may run into difficulties finding the desired number and type of landmarks with the desired spacing. This means that although maps keep improving, the work is slow and there is still a long way to go. It's almost as difficult (and arbitrary) to define a "complete" genome map as it is to define a "complete" genome sequence. One definition of "complete" is a map that includes the sequence and location of all of an organism's genes. Such maps currently exist for more than 150 organisms, most of them viruses with small genomes.

MAP BASED CLONING FOR GENE ISOLATION

Map based cloning can be used for isolation of a gene for which a mutation can be identified and in a crop for which a saturated molecular map is available. Availability of comprehensive YAC or BAC. Genomic library is another requirement. Most of these requirements are met in major crop plants. The mutation is first mapped close to any of the molecular markers that are already mapped. A variety of molecular markers are available for the preparation of these molecular maps (e.g., RFLP, RAPD, SSR, STMS, AFLP, SNP, etc.; consult next chapter for molecular markers).

Once the mutant is mapped, the most closely linked pair of flanking markers are used as hybridization probes to isolate clones containing the genomic segment located between the markers. Techniques of chromosome walking, chromosome jumping and chromosome landing can be used to reach closer to the gene of interest and to reach closer to the gene of interest and to isolated. These probable clones can be used for transformation of the mutant, to find out which of the clones complements the mutation. This clone then should contain the gene. Using this approach, several genes have been isolated in crop plants and animals including humans. A gene for cystic fibrosis in humans was cloned through map based cloning utilizing the technique will be increasingly used in future of isolation of a variety of genes for which the gene products are not known.

Simple sequence repeats (SSRs).

SSRs, also known as short tandem repeats (STRs) or microsatellites (1-6 bases long), are ubiquitous in eukaryotic genomes and can be analyzed through PCR technology. The sequences flanking specific microsatellite loci in a genome are believed to be conserved within a particular species, across species within a genus and rarely even across related genera. These flanking sequences, therefore, have been used to design primers for individual microsatellite loci (for details, see later) and the technique is described as sequence tagged micro satellite site (STMS) analysis or as simple sequence length polymorphism (SSLP).

The STMS or SSR markers reveal polymorphisms due to variation in the lengths of microsatellites at specific individual loci; they are, therefore, poly allelic and co-dominant in nature, thus proving to be very useful. Consequently, they have been used extensively not only for mapping SSR loci in human, mouse and many crop plants, but also for tagging genes in a variety of organisms.

Since development of SSR markers requires cloning and sequencing, initially it is very costly and labour-intensive, but once the locus specific primers become available, the approach becomes cost-effective. STMS primers, based on sequences flanking individual microsatellites, can be developed either through a search for microsatellites in the DNA sequence databases (including genomic sequences and cDNA sequences or ESTs) or through sequencing of restriction fragments or clones carrying microsatellites. Using the above methods, STMS primers have now become available not only in human and mouse but also in several crops including rice and bread wheat.

However, more recently emphasis has also been laid on EST-derived SSRs, since they represent the transcribed part of the genome and therefore may have higher level of transferability. In SSR analysis, high resolution, even without applying radioactivity, can be achieved through the use of polyacrylamide gels in combination with either ethidium bromide staining or silver staining.

The time needed for SSR analysis may be reduced by using one or more of the following approaches:

(i) High throughput approaches of DNA extraction (since high quality DNA is not needed for PCR).

(ii) Multiplexing facilitated either by size differences in the amplified products or through the use of fluorescent primers.

(iii) Multiple loading in a series of tiers on the same slab gel.

(iv) Laser detection through the use of automated DNA sequencers.

(v) Use of capillary electrophoresis (CE) in place of conventional slab gel electrophoresis.

In situ HYBRIDIZATION

In situ hybridization, as the name suggests, is a method of localizing and detecting specific mRNA sequences in morphologically preserved tissues sections or cell preparations by hybridizing the complementary strand of a nucleotide probe to the sequence of interest. Normal hybridization requires the isolation of DNA or RNA, separating it on a gel, blotting it onto nitrocellulose and probing it with a complementary sequence. The basic principles for *in situ* hybridization are the same, except one is utilizing the probe to detect specific nucleotide sequences within cells and tissues. The sensitivity of the technique is such that threshold levels of detection are in the region of 10-20 copies of mRNA per cell.

In situ hybridization presents a unique set of problems as the sequence to be detected will be at a lower concentration, be masked because of associated protein, or protected within a cell or cellular structure. Therefore, in order to probe the tissue or cells of interest one has to increase the permeability of the cell and the visibility of the nucleotide sequence to the probe without destroying the structural integrity of the cell or tissue. There are almost as many methods for carrying out *in situ* hybridization as there are tissues that have been probed. So more important than having a recipe is to have an understanding of the different stages in the process and their purpose.

Tissue

The most common tissue sections used with *in situ* hybridization are:

Frozen sections
Fresh tissue is snap frozen (rapidly put into a -80 freezer) and then when frozen embedded in a special support medium for thin cryosectioning. The sections are lightly and rapidly fixed in 4% paraformaldehyde just prior to processing for hybridization.

Paraffin embedded sections
Sections are fixed in formalin as one would normally fix tissues for histology and then embedded in wax (paraffin sections) before being sectioned.

Cells in suspension
Cells can be cytospun onto glass slides and fixed with methanol.

Choice of Probe

Probes are complimentary sequences of nucleotide bases to the specific mRNA sequence of interest. These probes can be as small as 20-40 base pairs or be up to 1000 bp. Although ultimately the question you ask and the type of sequence you are trying to detect is the overriding factor, one needs to optimize, as much as possible the conditions one uses. The strength of the bonds between the probe and the target plays an important role. The strength decreases in the order RNA-RNA to DNA-RNA. This stability is in turn influenced by the various hybridization conditions such as concentration of formamide, salt concentration, hybridization temperature, and pH. There are essentially four types of probe that can be used in performing in situ hybridization.

Oligonucleotide probes
These are produced synthetically by an automated chemical synthesis. The method utilizes readily available deoxynucleotides which are economical, but of course requires the specific nucleotide sequence to prepare. Designing the

sequence of the probe is one of the more critical decisions required when using oligonucleotide probes and is just not a matter of picking any region within the coding region of the target gene to bind to but requires careful design taking into account a number of issues (read below). These probes have the advantages of being resistant to RNases and are small, generally around 40-50 base-pairs. This is ideal for *in situ* hybridization because their small size allows for easy penetration into the cells or tissue of interest. In addition, because they are synthetically designed, it is possible to make a series of probes that have the same GC content; Since G/C base pairs bond more strongly than A/U base pairs, differences in GC content would require different hybridization conditions, so with oligonucleotides protocols can be standardized for many different probes irrespective of the target genes being measured. Another advantage of the oligonucleotide probes is that they are single stranded therefore excluding the possibility of renaturation.

Single Stranded DNA Probes

These have similar advantages to the oligonucleotide probes except they are much larger, probably in the 200-500 bp size range. They can be produced by reverse transcription of RNA or by amplified primer extension of a PCR-generated fragment in the presence of a single antisense primer. That is, once you have amplified the sequence of interest, a subsequent round of PCR is carried out using the first PCR product as template, but only using the anti-sense primers, thus producing single stranded DNA. This is therefore their disadvantage. They require time to prepare, expensive reagents are used during their preparation and good repertoires of molecular skills are required for their use.

Double Stranded DNA Probes

These can be produced by the inclusion of the sequence of interest in a bacteria which is replicated, lysed and the DNA extracted, purified and the sequence of interest is excised with restriction enzymes. On the other hand, if the sequence is known then by designing appropriate primers one can produce the relevant sequence very rapidly by PCR, potentially obtaining a very clean sample. The advantage of the bacterial preparation is that it is possible to obtain large quantities of the probe sequence in question. Because the probe is double stranded, it means that denaturation or melting has to be carried out prior to hybridization in order for one strand to hybridize with the mRNA of interest. These probes are generally less sensitive because of the tendency of the DNA strands to rehybridize to each other and are not as widely used today.

RNA Probes

RNA probes have the advantage that RNA-RNA hybrids are very thermostable and are resistant to digestion by RNases. This allows the possibility of post-hybridization digestion with RNase to remove non-hybridized RNA and therefore reduces the possibility of background staining. There are two methods of preparing RNA probes:

1) Complimentary RNA's are prepared by an RNA polymerase-catalyzed transcription of mRNA in the 3' to 5' prime direction.

2) Alternatively, in vitro transcription of linearized plasmid DNA with RNA polymerase can be used to produce the RNA probes. Here plasmid vectors containing polymerase from bacteriophages T3, T7 or SP6 are used

These probes however can be very difficult to work with as they are very sensitive to RNases (ubiquitous RNA degrading enzymes) and so scrupulous sterile technique should be observed or these probes can easily be destroyed. So saying this, RNA probes are still probably the most widely used probes with in situ hybridization. Oligonucleotide gene probes have multiple advantages over RNA or cDNA probes when used for in situ hybridization. Stability, Availability, Faster and less expensive to use, Easier to work with, More specific, Better tissue penetration, Better reproducibility and a wide range of labeling methods that do not interfere with target detection.

FLUORESCENCE IN SITU HYBRIDIZATION (FISH)

Fluorescence in situ hybridization (FISH), the assay of choice for localization of specific nucleic acids sequences in native context, is a 20-year-old technology that has developed continuously. Over its maturation, various methodologies and modifications have been introduced to optimize the detection of DNA and RNA. The pervasiveness of this technique is largely because of its wide variety of applications and the relative ease of implementation and performance of in situ studies. Although the basic principles of FISH have remained unchanged, high sensitivity detection, simultaneous assay of multiple species, and automated data collection and analysis have advanced the field significantly. The introduction of FISH surpassed previously available technology to become a foremost biological assay. Key methodological advances have allowed facile preparation of low-noise hybridization probes, and technological breakthroughs now permit multi-target visualization and quantitative analysis – both factors that have made FISH accessible to all and applicable to any investigation of nucleic acids. In the future, this technique is likely to have significant further impact on live-cell imaging and on medical diagnostics.

Fluorescent in situ hybridization (FISH) is a powerful technique for detecting RNA or DNA sequences in cells, tissues, and tumors. FISH provides a unique link among the studies of cell biology, cytogenetics, and molecular genetics.

Fluorescent in situ hybridization is a technique in which single-stranded nucleic acids (usually DNA, but RNA may also be used) are permitted to interact so that complexes, or hybrids, are formed by molecules with sufficiently similar, complementary sequences. Through nucleic acid hybridization, the degree of sequence identity can be determined, and specific sequences can be detected and located on a given chromosome. The method comprises of three basic steps: fixation of a specimen on a microscope slide, hybridization of labeled probe to homologous fragments of genomic DNA, and enzymatic detection of the tagged target hybrids. While probe sequences were initially detected with isotopic reagents, nonisotopic hybridization has become increasingly popular, with fluorescent hybridization now a common choice. Protocols involving nonisotopic probes are considerably faster, with greater signal resolution, and provide options to visualize different targets simultaneously by combining various detection methods(Fig 13.1).

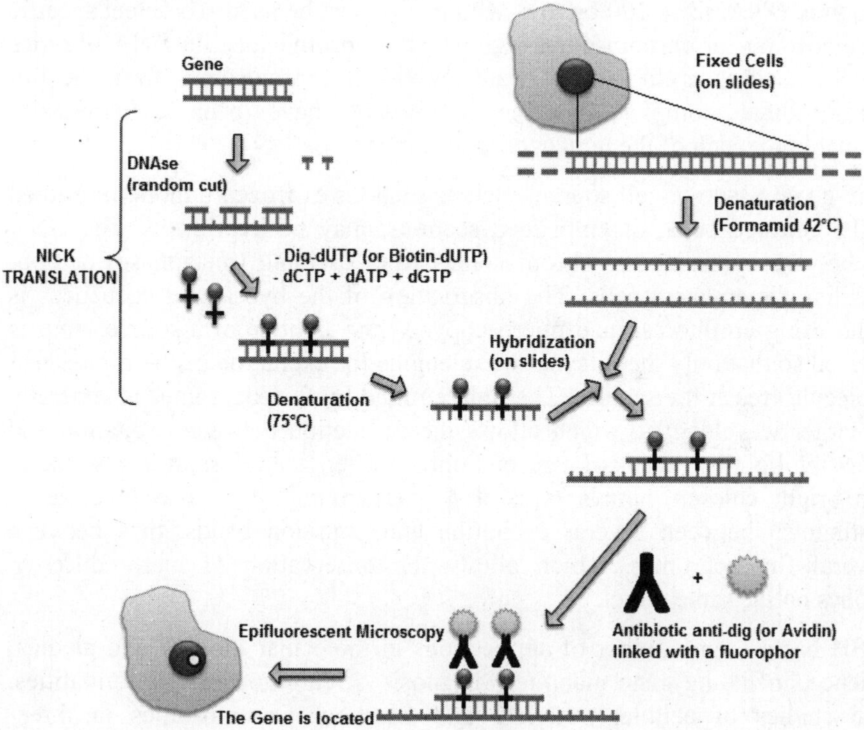

Fig 13.1: FISH

The detection of sequences on the target chromosomes is performed indirectly, commonly with biotinylated or digoxigenin-labeled probes detected via a fluorochrome-conjugated detection reagent, such as an antibody conjugated with fluorescein. As a result, the direct visualization of the relative position of the probes is possible. Increasingly, nucleic acid probes labeled directly with fluorochromes are used for the detection of large target sequences. This method takes less time and results in lower background; however, lower signal intensity is generated. Higher sensitivity can be obtained by building layers of detection reagents, resulting in amplification of the signal. Using such means, it is possible to detect single-copy sequences on chromosome with probes shorter than 0.8 kb.

Probes can vary in length from a few base pairs for synthetic oligonucleotides to larger than one Mbp. Probes of different types can be used to detect distinct DNA types. PCR-amplified repeated DNA sequences, oligonucleotides specific for repeat elements, or cloned repeat elements can be used to detect clusters of repetitive DNA in heterochromatin blocks or centromeric regions of individual chromosomes. These are useful in determining aberrations in the number of chromosomes present in a cell. In contrast, for detecting single locus targets, cDNAs or pieces of cloned genomic DNA, from 100 bp to 1 Mbp in size, can be used. To detect specific chromosomes or chromosomal regions, chromosome-specific DNA libraries can be used as probes to delineate individual chromosomes from the full chromosomal complement. Specific probes have been commercially available for each of the human chromosomes since 1991.

Any given tissue or cell source, such as sections of frozen tumors, imprinted cells, cultured cells, or embedded sections, may be hybridized. The DNA probes are hybridized to chromosomes from dividing (metaphase) or non-dividing (interphase) cells. The observation of the hybridized sequences is done using epifluorescence microscopy. White light from a source lamp is filtered so that only the relevant wavelengths for excitation of the fluorescent molecules reach the sample. The light emitted by fluorochromes is generally of larger wavelengths, which allows the distinction between excitation and emission light by means of a second optical filter. Therefore, it is possible to see bright colored signals on a dark background. It is also possible to distinguish between several excitation and emission bands, thus between several fluorochromes, which allows the observation of many different probes on the same target.

FISH has a large number of applications in molecular biology and medical science, including gene mapping, diagnosis of chromosomal abnormalities, and studies of cellular structure and function. Chromosomes in three-dimensionally preserved nuclei can be "painted" using FISH. In clinical research FISH can be used for prenatal diagnosis of inherited chromosomal aberrations, postnatal diagnosis of carriers of genetic disease, diagnosis of infectious disease, viral and bacterial disease, tumor cytogenetic diagnosis, and detection of aberrant gene expression. In laboratory research, FISH can be used for mapping chromosomal genes, to study the evolution of genomes (Zoo FISH), analyzing nuclear organization, visualization of chromosomal territories and chromatin in interphase cells, to analyze dynamic nuclear processes, somatic hybrid cells, replication, chromosome sorting, and to study tumor biology. It can also be used in developmental biology to study the temporal expression of genes during differentiation and development. Recently, high resolution FISH has become a popular method for ordering genes or DNA markers within chromosomal regions of interest.

CHROMOSOME MICRODISSECTION AND MICROCLONING

The technique of chromosome microdissection and microcloning has been developed for more than 20 years. As a bridge between cytogenetics and molecular genetics, it leads to a number of applications: chromosome painting probe isolation, genetic linkage map and physical map construction, and expressed sequence tags generation. During those 20 years, this technique has not only been benefited from other technological advances but also cross-fertilized with other techniques. Today, it becomes a practicality with extensive uses.

History

Scalenghe *et al.* (1981) were the first to develop the chromosome microdissection and microcloning technique. They used this technique to generate DNA from individual bands of *Drosophila melanogaster* polytene chromosomes and obtained 80 clones. This technique was then applied to mouse to obtain 212 microclones from the proximal half of chromosome 17 containing the t-complex. Then, it was extended to human chromosomes. However, at that time, studies were mainly focusing on some chromosomes that are easily identifiable by their configuration, such as the X chromosome of mouse and chromosome 2 of human. After the advent of G-banding technique, which makes the identification of human and animal chromosomes easier, and PCR technique, the chromosome microdissection and microcloning technique was extensively used in human and animal genomics research.

To microdissect chromosomes in plant is more difficult than in human, because chromosome preparation is more difficult in plant. The chromosome microdissection and mi-crocloning technique was applied to isolate B-chromosome DNA from rye in 1991. It was the first case that this method was used on plant chromosomes. However, work has been limited to chromosomes that are easy to identify, such as the satellite chromosomes, the largest or the smallest chromosome, and B chromosome. Chromosomes in cytogenetic stocks that can be easily distinguished were used for chromosome microdissection, such as telo-chromosomes, addition lines, and reconstructed translocation chromosomes.

Benefits from other Techniques

Chromosome microdissection and microcloning has been benefited from technological advances and coupling with other techniques, which further improved its application.

In the microdissection and microcloning technique, the chromosome was initially dissected with glass microneedles under an inverted microscopy. Even for an expert, it is difficult to dissect and collect a large numbers of chromosome fragments from the same region. Monajembashi *et al.* (1986) developed a method to dissect target chromosomes by using laser microbeam. The equipment they used is mainly composed of argon-ion laser power supply, microcomputer and an invert microscope. Even though the intensity and the position of the laser beam were controlled by a microcomputer, the collection of the target region is still a difficult step for the operators. These two methods were combined together. Firstly, the laser beam was used to dissect the targeted region of chromosomes, and then, the glass microneedle was used to collect the targeted regions. Flow cytometry (FCM) has been applied to the vast field of cytogenetics research through adaptation to the observation of isolated chromosomes since 1975. Flow cytometry, while successful in isolating some of the larger chromosomes was clearly limited in its sensitivity for isolating the smaller ones. In order to isolate one special chromosome, Griffin *et al.* (1999) combined the two approaches (flow cytometry and microdissection) and successfully collected the targeted chromosome of chicken. From glass microneedle to laser microbeam, and then flow cytometric technology, chromosome microdissection method had undergone changes from manual operation to computer driven manipulation. As a result, both the rate and precision of chromosome isolation are improved. Moreover, these dissection techniques have complementary advantages.

The earliest chromosome microcloning was a kind of direct cloning of the dissected chromosomal material in a nanoliter microdrop contained in an oil chamber. A very large number of dissected chromosome fragments had to be used and a relatively low cloning efficiency was attained. However, the introduction of PCR technology to the micro-cloning protocols has brought a substantial improvement in cloning efficiency. From that on, the number of dissected chromosome fragments needed was significantly reduced, and two new microcloning methods were developed based on PCR technology:

1) Adaptor mediated PCR (LA-PCR).
2) Degenerated oligonuleotide-primed-PCR (DOP-PCR).

With the LA-PCR method, all the enzymatic manipulations of the dissected chromosomes, including PCR, can be easily performed in a single 0.5 ml tube by simply adding the appropriate components for each subsequent step.

It avoids the complex micromanipulation in a microchamber and can generate much larger fragments—the average length is about 300-2,000 bps. Using this method, Chen and Arm-strong (1995) constructed a single chromosome (less than 0.4 pg) library, potentially comprising of 500,000 recombinant clones.

The DOP-PCR technique, which is rapid, efficient, and species-independent, is designed to amplify target DNAs at frequently occurring priming sites using the primer of partially degenerate sequence, without restrictions imposed by the complexity or the origin of DNA. As an important method of microcloning, DOP-PCR overcomes the problems of regional bias and species dependence seen within LA-PCR. It is a simple PCR technique involving multiple loci priming, which allows a more general amplification than LA-PCR.

The combination of improved micromanipulation methods and PCR technology has enabled scientists to dissect specific chromosomes or chromosomal regions both accurately and frequently, thus, improving the efficiency of this technique.

The Applications of Chromosome Microdissection and Microcloning in Genomic Research

Chromosomal microdissection and microcloning provides a direct approach for isolating DNA from any cytogenetically recognizable region of a chromosome. The isolated DNA can be used for genomic research including:

1) Genetic linkage map and physical map construction.
2) Generation of probes for chromosome painting.
3) Generation of chromosome specific expressed sequence tags libraries.

Providing Probes for Genetic Linkage Map and Physical Map Construction

Despite the rapid progress in gene mapping in recent years, there are still large areas of the genome for which markers are sparse or which are completely unmapped. Direct chromosomal microdissection and microcloning is a rapid technique for providing probes for such areas. The clones of specific chromosomes libraries can be used in conjunction with existing markers to construct a fine genetic linkage map and physical map, providing a gateway for understanding of chromosome structure and organization of a specific region of the genome. Combined with chromosome walking, chromosome region-specific physical map can be constructed by using chromosomes or chromosome fragments library to select cosmid library yeast artificial chromosome (YAC) and bacterial artificial chromosome (BAC). Moreover, microdissect DNA clones can be transformed to sequence tagging sites (STS), then the STSs are used to select cosmid library, YAC or BAC for eventually drawing a physical map.

Generation of Chromosome Painting Probes

Fluorescence in situ hybridization (FISH) plays an essential role in research and clinical diagnostics. The versatility and resolution of FISH depends critically on the probe set used. The probes usually generate from the clones of cosmid library, yeast artificial chromosome (YAC) and bacterial artificial chromosome (BAC). Microdissect DNA clones are also the source of chromosome painting probes, particularly in the research of human genome. The technique that combines chromosome microdissection and chromosome painting is named micro-FISH. As one of the important applications of chromosome microdissection, micro-FISH can be used to identify the reliability of the origin of microdissect DNA. On the other hand, micro-FISH is an important tool for other research, such as chromosome construction aberration, chromosome origin identification and comparative analysis of genomes. More and more chromosomal sites in the human genome have been identified as primary lesions in specific genetic diseases or cancers, such as the 4p16.1 for Huntington's disease, 22q11 for cat eye syndrome, 3p14-p23 in small-cell lung cancer and so on. Isolating the sequences from human chromosomal regions associated with specific genetic diseases or cancers to understand disease related genes is the prevalent approach in latest clinical research. Arens *et al.* (2004) micro dissected the GTG band of the aberrant chromosome as the chromosome painting probes that identified the inter-translocation and the breakpoint of the chromosome of a large family with an unbalanced insertion translocation (3; 5)(q25.3; q22.1q31.3). Vermeesch *et al.* (2005) showed the feasibility of using micro dissected chromosomes or chromosomal fragments to molecularly map the chromosomal breakpoints on array Comparative Genomic Hybridization (CGH). To study the behavior of genes conferring drug resistance of cancer lines, Mahjoubi *et al.* (2006) isolated the genes residing in a homogeneously staining region (HSR) in drug-resistant cell sublines. The isolated sequence provided a resource for future investigations in searching for novel genes contributing to drug resistance. 'Chromosome painting' in plants is relatively underde-veloped, because of the large genomes and large quantity of homologous sequences shared among chromosomes. Using micro dissected plant chromosome DNA, there are many advantages to paint chromosomes from various plant species with large genomes, such as barley, wheat, *Vi-cia faba*, and *Secale cereale*. However, no specific paintings of chromosomes were observed. The painting probes were obtained only from B chromosomes and Y chromosome, because of the abundance of chromosome specific repetitive sequences.

Generating ESTs of Specific Chromosomes

Expressed sequence tags (ESTs) are short (200-500) base pair "single pass" sequencing reads derived from both 3' and 5' end of cDNA clones selected at random. The ESTs give important information about its coding content and expression patterns in different tissues and organs of organisms at different developmental stages and environments. That is why partial cDNA sequencing to generate ESTs is being used at present for the fast and efficient profiling of gene expression in various tissues, cell types, or developmental stages. If the ESTs could be isolated from those specific chromosomes and specific chromosome regions directly, it would be easier to study and isolate those genes. Analysis of ESTs constitutes a useful approach for gene identification and gaining ESTs from cDNA libraries is the normal method. However, the dataset is huge and highly redundant, and it is difficult to identify the chromosome specific ESTs from cDNA libraries. Several methods to isolate chromosome specific and chromosome region specific expressed sequences had been developed, mainly including (1) restriction enzyme mapping of CpG islands; (2) exon trapping; (3) direct selection; (4) Hybrid selection and (5) Microdissection-mediated cDNA capture. First two of them, however, are labor-intensive and technically complex.

MOLECULAR MARKERS AND GENOME ANALYSIS

With the advent of molecular markers, a new generation of markers has been introduced over the last two decades, which has revolutionized the entire scenario of biological sciences. DNA-based molecular markers have acted as versatile tools and have found their own position in various fields like taxonomy, physiology, embryology, genetic engineering, etc. They are no longer looked upon as simple DNA fingerprinting markers in variability studies or as mere forensic tools. Ever since their development, they are constantly being modified to enhance their utility and to bring about automation in the process of genome analysis. The discovery of PCR (polymerase chain reaction) was a landmark in this effort and proved to be an unique process that brought about a new class of DNA profiling markers. This facilitated the development of marker-based gene tags, map-based cloning of agronomically important genes, variability studies, phylogenetic analysis, synteny mapping, marker-assisted selection of desirable genotypes, etc. Thus giving new dimensions to concerted efforts of breeding and marker-aided selection that can reduce the time span of developing new and better varieties and will make the dream of super varieties come true. These DNA markers offer several advantages over traditional phenotypic markers, as they provide data that can be analyzed objectively.

DNA-based molecular markers

Genetic polymorphism is classically defined as the simultaneous occurrence of a trait in the same population of two or more discontinuous variants or genotypes. Although DNA sequencing is a straightforward approach for identifying variations at a locus, it is expensive and laborious. A wide variety of techniques have, therefore, been developed in the past few years for visualizing DNA sequence polymorphism.

The term DNA-fingerprinting was introduced for the first time by Alec Jeffrey in 1985 to describe bar-code-like DNA fragment patterns generated by multilocus probes after electrophoretic separation of genomic DNA fragments. The emerging patterns make up a unique feature of the analysed individual and are currently considered to be the ultimate tool for biological individualization. Recently, the terms DNA fingerprinting/profiling is used to describe the combined use of several single locus detection systems and are being used as versatile tools for investigating various aspects of plant genomes. These include characterization of genetic variability, genome fingerprinting, genome mapping, gene localization, analysis of genome evolution, population genetics, taxonomy, plant breeding, and diagnostics.

Properties desirable for ideal DNA markers

- Highly polymorphic nature.
- Codominant inheritance (determination of homozygous and heterozygous states of diploid organisms).
- Frequent occurrence in genome.
- Selective neutral behaviour (the DNA sequences of any organism are neutral to environmental conditions or management practices).
- Easy access (availability).
- Easy and fast assay.
- High reproducibility.
- Easy exchange of data between laboratories.

It is extremely difficult to find a molecular marker which would meet all the above criteria. Depending on the type of study to be undertaken, a marker system can be identified that would fulfill at-least a few of the above characteristics.

Various types of molecular markers are utilized to evaluate DNA polymorphism and are generally classified as hybridization-based markers and polymerase chain reaction (PCR)-based markers. In the former, DNA profiles are visualized by hybridizing the restriction enzyme-digested DNA, to a labelled probe, which is a DNA fragment of known origin or sequence. PCR-based markers involve *in vitro* amplification of particular DNA sequences or loci, with the help of specifically or arbitrarily chosen oligonucleotide sequences (primers) and a thermostable DNA polymerase enzyme. The amplified fragments are separated electrophoretically and banding patterns are detected by different methods such as staining and autoradiography. PCR is a versatile technique invented during the mid-1980s. Ever since thermostable DNA polymerase was introduced in 1988, the use of PCR in research and clinical laboratories has increased tremendously. The primer sequences are chosen to allow base-specific binding to the template in reverse orientation. PCR is extremely sensitive and operates at a very high speed. Its application for diverse purposes has opened up a multitude of new possibilities in the field of molecular biology.

Types and description of DNA markers

Different DNA Markers

AFLP - amplified fragment length polymorphism

AMP-PCR- anchored microsatellite primed PCR

AP-PCR - arbitrarily primed PCR

ASA- allele-specific amplification

ASSR - anchored simple sequence repeat

CAPS - cleaved amplified polymorphic sequence

DAF - DNA amplification fingerprint

DALP - direct amplification of length polymorphism

DAMD-PCR - direct amplification of microsatellite DNA by PCR

DFLP - DNA fragment length polymorphism

dRAMP - digested RAMP

IFLP - intron-retrotransposon amplified polymorphism

IM-PCR - inter-microsatellite PCR

IRAP - inter-retrotransposon amplified polymorphism

ISA - inter-SSR amplification

ISSR - inter-simple sequence repeats

MAAP - multiple arbitrary amplicon profiling

MP-PCR - microsatellite-primed PCR

OLA - oligonucleotide ligation assay

RAHM - randomly amplified hybridizing microsatellites

RAMPO - randomly amplified microsatellite polymorphism

RAMP - randomly amplified microsatellites polymorphism

RAMS - randomly amplified microsatellites

RAPD - random amplified polymorphic DNA

RBIP - retrotransposon-based insertion polymorphism

REMAP - retrotransposon-microsatellite amplified polymorphism

RFLP - restriction fragment length polymorphism

SAMPLE - selective amplification of microsatellite polymorphic loci

SCAR - sequence characterized amplified regions

SNP - single nucleotide polymorphism

SPAR - single primer amplification polymorphism

S-SAP - sequence-specific amplification polymorphism

SSCP - single strand conformation polymorphism

SSLP - simple sequence length polymorphism

SSR - simple sequence repeat

STAR - sequence tagged microsatellite region

STMS - sequence-tagged microsatellite site

STR - short tandem repeat

STS - sequence-tagged-site

VNTR - variable number of tandem repeats

Restriction fragment length polymorphism (RFLP)

RFLPs are simply inherited naturally occurring Mendelian characters. They have their origin in the DNA rearrangements that occur due to evolutionary processes, point mutations within the restriction enzyme recognition site sequences, insertions or deletions within the fragments, and unequal crossing over.

In RFLP analysis, restriction enzyme-digested genomic DNA is resolved by gel electrophoresis and then blotted on to a nitrocellulose membrane. Specific banding patterns are then visualized by hybridization with labelled probe. These probes are mostly species-specific single locus probes of about 0.5–3.0 kb in size, obtained from a cDNA library or a genomic library. RFLPs, being co dominant markers, can detect coupling phase of DNA molecules, as DNA fragments from all homologous chromosomes are detected. They are very reliable markers in linkage analysis and breeding and can easily determine if a linked trait is present in a homozygous or heterozygous state in individual, information highly desirable for recessive traits. However, their utility has been hampered due to the large amount of

DNA required for restriction digestion and Southern blotting. The requirement of radioactive isotope makes the analysis relatively expensive and hazardous. The assay is time-consuming and labor-intensive and only one out of several markers may be polymorphic, which is highly inconvenient especially for crosses between closely-related species. Their inability to detect single base changes restricts their use in detecting point mutations occurring within the regions at which they are detecting polymorphism.

RFLP markers converted in to PCR based-markers

Sequence-tagged sites (STS): RFLP probes specifically linked to a desired trait can be converted into PCR-based STS markers based on nucleotide sequence of the probe giving polymorphic band pattern, to obtain specific amplicon. Using this technique, tedious hybridization procedures involved in RFLP analysis can be overcome. This approach is extremely useful for studying the relationship between various species. When these markers are linked to some specific traits, for example powdery mildew resistance gene or stem rust resistance gene in barley, they can be easily integrated into plant breeding programmes for marker-assisted selection of the trait of interest.

Allele-specific associated primers (ASAPs): To obtain an allele-specific marker, specific allele (either in homozygous or heterozygous state) is sequenced and specific primers are designed for amplification of DNA template to generate a single fragment at stringent annealing temperatures. These markers tag specific alleles in the genome and are more or less similar to SCARs.

Expressed sequence tag markers (EST): This term was introduced by Adams *et al.* Such markers are obtained by partial sequencing of random cDNA clones. Once generated, they are useful in cloning specific genes of interest and synteny mapping of functional genes in various related organisms. ESTs are popularly used in full genome sequencing and mapping programmes underway for a number of organisms and for identifying active genes thus helping in identification of diagnostic markers. Moreover, an EST that appears to be unique helps to isolate new genes. EST markers are identified to a large extent for rice, *Arabidopsis*, etc. wherein thousands of functional cDNA clones are being converted in to EST markers.

Single strand conformation polymorphism (SSCP): This is a powerful and rapid technique for gene analysis particularly for detection of point mutations and typing of DNA polymorphism. SSCP can identify heterozygosity of DNA fragments of the same molecular weight and can even detect changes of a few nucleotide bases as the mobility of the single-stranded DNA changes with change in its GC content due to its conformational change. To overcome problems of reannealing and complex banding patterns, an improved technique called asymmetric-PCR SSCP was developed, wherein the denaturation step was eliminated and a large-sized sample could be loaded for gel electrophoresis, making it a potential tool for high throughput DNA polymorphism. It was found useful in the detection of heritable human diseases. In plants, however, it is not well developed although its application in discriminating progenies can be exploited, once suitable primers are designed for agronomically important traits.

Dispersed repetitive DNA (drDNA): Nuclear genomes of all eukaryotes contain repeats of short sequence motifs (2-15 bp in length), dispersed throughout the genome, which have been described either as dispersed repetitive DNA (drDNA), or as variable number of tandem repeats (VNTRs). These sequences mainly include minisatellites and microsatellites {the microsatellites have also been described as simple sequence repeats (SSRs) or short tandem repeats (STRs) and can be detected, either through nucleic acid hybridization or through PCR amplification. In the hybridization approach, cloned or synthetic oligonucleotides representing any of these short repeats are utilized as probes. Polymorphism in drDNA is resolved by RFLP technology, which is sometimes modified through the use of in-gel hybridization and, therefore, described as oligonucleotide in-gel hybridization/fingerprinting (in this in-gel technology, electrophoresed

DNA fragments are not transferred to a filter as dome in Southern blotting for RFLPs, but instead the gels are dried and used directly for hybridization).

These repeat sequences have been used as multi-locus probes, thus having the advantage of revealing polymorphisms at many loci simultaneously. However, in some cases, as observed in tomato and bread wheat, they give only a few bands thus limiting their utility. Another disadvantage often attributed to the technique of in-gel hybridization is, that perhaps it does not allow detection of small segments representing unique sequences carrying these dispersed repeat sequences.

Repetitive DNA

A major step forward in genetic identification is the discovery that about 30–90% of the genome of virtually all the species is constituted by regions of repetitive DNA, which are highly polymorphic in nature. These regions contain genetic loci comprising several hundred alleles, differing from each other with respect to length, sequence or both and they are interspersed in tandem arrays ubiquitously. The repetitive DNA regions play an important role in absorbing mutations in the genome. Of the mutations that occur in the genome, only inherited mutations play a vital role in evolution or polymorphism. Thus repetitive DNA and mutational forces functional in nature together form the basis of a number of marker systems that are useful for various applications in plant genome analysis. The markers belonging to this class are both hybridization-based and PCR-based.

Microsatellites and Minisatellites

The term microsatellites were coined by Litt and Lutty, while the term minisatellites was introduced by Jeffrey. Both are multilocus probes creating complex banding patterns and are usually non-species specific occurring ubiquitously. They essentially belong to the repetitive DNA family. Fingerprints generated by these probes are also known as oligonucleotide fingerprints. The methodology has been derived from RFLP and specific fragments are visualized by hybridization with a labelled micro- or minisatellite probe.

Minisatellites are tandem repeats with a monomer repeat length of about 11–60 bp, while microsatellites or short tandem repeats/simple sequence repeats (STRs/ SSRs) consist of 1 to 6 bp long monomer sequence that is repeated several times. These loci contain tandem repeats that vary in the number of repeat units between genotypes and are referred to as variable number of tandem repeats (VNTRs) (i.e. a single locus that contains variable number of tandem repeats between individuals) or hypervariable regions (HVRs) (i.e. numerous loci containing tandem repeats within a genome generating high levels of polymorphism between individuals). Microsatellites and minisatellites thus form an ideal marker system creating complex banding patterns by simultaneously detecting multiple DNA loci. Some of the prominent features of these markers are that they are dominant fingerprinting markers and codominant STMS (sequence tagged microsatellites) markers. Many alleles exist in a population, the level of heterozygosity is high and they follow Mendelian inheritance.

Minisatellite and Microsatellite sequences converted into PCR-based markers

Sequence-tagged microsatellite site markers (STMS): This method includes DNA polymorphism using specific primers designed from the sequence data of a specific locus. Primers complementary to the flanking regions of the simple sequence repeat loci yield highly polymorphic amplification products. Polymorphisms appear because of variation in the number of tandem repeats (VNTR loci) in a given repeat motif. Tri- and tetranucleotide microsatellites are more popular for STMS analysis because they present a clear banding pattern after PCR and gel electrophoresis. However, dinucleotides are generally abundant in genomes and have been used as markers e.g. (CA)*n*(AG)*n* and (AT)*n*. The di- and tetranucleotide repeats are present mostly in the non-coding regions of the genome, while 57% of trinucleotide repeats are shown to reside in or around the genes. A very good relationship between the number of alleles detected and the total number of simple repeats within the targeted microsatellite DNA has been observed. Thus larger the repeat number in the microsatellite DNA, greater is the number of alleles detected in a large population.

Arbitrary sequence markers

Randomly-amplified polymorphic DNA markers (RAPD)

In 1991 Welsh and McClelland developed a new PCR-based genetic assay namely randomly amplified polymorphic DNA (RAPD). This procedure detects nucleotide sequence polymorphisms in DNA by using a single primer of arbitrary nucleotide sequence. In this reaction, a single species of primer anneals to the genomic DNA at two different sites on complementary strands of DNA template. If these priming sites are within an amplifiable range of each other, a discrete DNA product is formed through thermocyclic amplification. On an average, each primer directs amplification of several discrete loci in the genome, making the assay useful for efficient screening of nucleotide sequence polymorphism between individuals. However, due to the stoichastic nature of DNA amplification with random sequence primers, it is important to optimize and maintain consistent reaction conditions for reproducible DNA amplification. They are dominant markers and hence have limitations in their use as markers for mapping, which can be overcome to some extent by selecting those markers that are linked in coupling. RAPD assay has been used by several groups as efficient tools for identification of markers linked to agronomically important traits, which are introgressed during the development of near isogenic lines. The application of RAPDs and their related modified markers in variability analysis and individual-specific genotyping has largely been carried out, but is less popular due to problems such as poor reproducibility faint or fuzzy products, and difficulty in scoring bands, which lead to inappropriate inferences. Some variations in the RAPD technique include

DNA amplification fingerprinting (DAF)

Caetano-Anolles *et al.* employed single arbitrary primers as short as 5 bases to amplify DNA using polymerase chain reaction. In a spectrum of products obtained, simple patterns are useful as genetic markers for mapping, while more complex patterns are useful for DNA fingerprinting. Band patterns are reproducible and can be analysed using polyacrylamide gel electrophoresis and silver staining. DAF requires careful optimization of parameters; however, it is extremely amenable to automation and fluorescent tagging of primers for early and easy determination of amplified products. DAF profiles can be tailored by employing various modifications such as predigesting of template. This technique has been useful in genetic typing and mapping.

Arbitrary primed polymerase chain reaction (AP-PCR)
This is a special case of RAPD, wherein discrete amplification patterns are generated by employing single primers of 10–50 bases in length in PCR of genomic DNA. In the first two cycles, annealing is under non-stringent conditions. The final products are structurally similar to RAPD products. Compared to DAF, this variant of RAPD is not very popular as it involves autoradiography. Recently, however, it has been simplified by separating the fragments on agarose gels and using ethidium bromide staining for visualization.

Sequence characterized amplified regions for amplification of specific band (SCAR)
Michelmore *et al.* and Martin *et al.* introduced this technique wherein the RAPD marker termini are sequenced and longer primers are designed (22–24 nucleotide bases long) for specific amplification of a particular locus. These are similar to STS markers in construction and application. The presence or absence of the band indicates variation in sequence. These are better reproducible than RAPDs. SCARs are usually dominant markers, however, some of them can be converted into codominant markers by digesting them with tetra cutting restriction enzymes and polymorphism can be deduced by either denaturing gel electrophoresis or SSCP. Compared to arbitrary primers, SCARs exhibit several advantages in mapping studies (codominant SCARs are informative for genetic mapping than dominant RAPDs), map-based cloning as they can be used to screen pooled genomic libraries by PCR, physical mapping, locus specificity, etc. SCARs also allow comparative mapping or homology studies among related species, thus making it an extremely adaptable concept in the near future.

Amplified fragment length polymorphism (AFLP)
A recent approach by Zabeau *et al.*, known as AFLP, is a technique based on the detection of genomic restriction fragments by PCR amplification and can be used for DNAs of any origin or complexity. The fingerprints are produced, without any prior knowledge of sequence, using a limited set of generic primers. The number of fragments detected in a single reaction can be 'tuned' by selection of specific primer sets. AFLP technique is reliable since stringent reaction conditions are used for primer annealing. This technique thus shows an ingenious combination of RFLP and PCR techniques and is extremely useful in detection of polymorphism between closely related genotypes. AFLP procedure mainly involves 3 steps:

1) Restriction of DNA using a rare cutting and a commonly cutting restriction enzyme simultaneously (such as *Mse*I and *Eco*RI) followed by ligation of oligonucleotide adapters, of defined sequences including the respective restriction enzyme sites.

2) Selective amplifications of sets of restriction fragments, using specifically designed primers. To achieve this, the 5' region of the primer is made such that it would contain both the restriction enzyme sites on either sides of the fragment complementary to the respective adapters, while the 3' ends extend for a few arbitrarily chosen nucleotides into the restriction fragments.
3) Gel analysis of the amplified fragments.

AFLP analysis depicts unique fingerprints regardless of the origin and complexity of the genome. Most AFLP fragments correspond to unique positions on the genome and hence can be exploited as landmarks in genetic and physical mapping. AFLPs are extremely useful as tools for DNA fingerprinting and also for cloning and mapping of variety-specific genomic DNA sequences. Similar to RAPDs, the bands of interest obtained by AFLP can be converted into SCARs. Thus AFLP provides a newly developed, important tool for a variety of applications.

14

Gene Therapy

Genes, which are carried on chromosomes, are the basic physical and functional units of heredity. Genes are specific sequences of bases that encode instructions on how to make proteins. Although genes get a lot of attention, it's the proteins that perform most life functions and even make up the majority of cellular structures. When genes are altered so that the encoded proteins are unable to carry out their normal functions, genetic disorders can result. Much attention has been focused on the so-called genetic metabolic diseases in which a defective gene causes an enzyme to be either absent or ineffective in catalyzing a particular metabolic reaction effectively. A potential approach to the treatment of genetic disorders in man is gene therapy. This is a technique whereby the absent or faulty gene is replaced by a working gene, so that the body can make the correct enzyme or protein and consequently eliminate the root cause of the disease.

Gene therapy is the introduction of genetic material into cells for therapeutic purposes. Recent scientific breakthroughs in the genomics field and our understanding of the important role of genes in disease has made gene therapy one of the most rapidly advancing fields of biotechnology with great promise for treating inherited and acquired diseases.

Many human diseases are caused by the absence or inappropriate presence of a protein. Biotechnology's first promise was to isolate and produce these natural proteins through genetic engineering and recombinant technology. The protein could then be administered to patients in order to compensate for its absence. Because proteins are not orally available, biotech companies focused on innovative methods of protein delivery and sustained drug delivery. Today, gene therapy is the ultimate method of protein delivery, in which the delivered gene enters the body's cells and turns them into small "factories" that produce a therapeutic protein for a specific disease over a prolonged period.

As gene therapy has moved from the laboratory into the clinic, several issues have emerged as central to the development of this technology: gene identification, gene expression and gene delivery. Gene identification was originally tackled by academic researchers supported by the government's

Human Genome Project and more recently through genomics companies. A number of disease-related genes with direct clinical have already been identified, and this number is growing as the field rapidly advances. Some of these genes are in the public domain and some are proprietary. Genes with broader clinical application are also being utilized to make cells express immune activating agents locally at the disease site or to become susceptible to further drug treatment or to immune response recognition.

Gene therapy is a technique for correcting defective genes responsible for disease development. Researchers may use one of several approaches for correcting faulty genes:

- A normal gene may be inserted into a nonspecific location within the genome to replace a nonfunctional gene. This approach is most common.
- An abnormal gene could be swapped for a normal gene through homologous recombination.
- The abnormal gene could be repaired through selective reverse mutation, which returns the gene to its normal function.
- The regulation (the degree to which a gene is turned on or off) of a particular gene could be altered.

Gene therapy promises to revolutionize medicine by treating the (genetic) causes of diseases rather than the symptoms. This radical improvement is possible because the gene based approach can provide superior targeting and prolonged duration of action. Moreover, gene therapy is a platform technology applicable to a wide range of diseases. Ongoing clinical studies are addressing a wide range of diseases and target cells, such as cardiovascular diseases, inherited monogenic disorders, rheumatoid arthritis (RA), cancer and cubal tunnel syndrome.

METHODS OF GENE TRANSFER

Gene therapy is broadly defined as the delivery or transfer of genetic material to target cells for therapeutic purposes. A current method of gene transfer includes the use of viral and nonviral vectors. Viral vectors accomplish the gene transfer directly by viral mediated infection. Non-viral gene transfer or transfection, can be attained through the chemical and physical treatment of the target cells. Criteria required to fulfill successful gene transfer includes the following:

1) A vehicle to deliver a therapeutic gene to the appropriate target cell.
2) An appropriate level of the expression of the therapeutic gene in the target tissue.

3) Most importantly, the transfer and expression of the therapeutic gene must not be deleterious for the patient or the environment.

Ex vivo Gene Transfer

This initially involves transfer of cloned genes into cells grown in culture. Those cells which have been transformed successfully are selected, expanded by cell culture in vitro, and then introduced in to the patient. To avoid immune system rejection of the introduced cells, autologous cells are normally used: the cells are collected initially from the patient to be treated and grown in culture before being reintroduced into the same individual(Fig. 14.1). Clearly, this approach is only applicable to tissues that can be removed from the body, altered genetically and returned to the patient where they will engraft and survive for a long period of time (e.g. cells of the hatematopoietic system and skin cells).

In vivo Gene Transfer

Here the cloned genes are transferred directly into the tissues of the patient. This may be the only possible option in tissues where individual cells cannot be cultured in vitro in sufficient numbers (e.g. brain cells) and /or where cultured cells cannot be re-implanted efficiently in patients. Liposomes and certain viral vectors are increasingly being employed for this purpose. In the latter case, it is often convenient to implant vector producing cells (VPCs), cultured cells which have been infected by the recombinant retrovirus in vitro: in this case the VPCs transfer the gene to the surrounding disease cells.

Classical gene therapies normally require efficient transfer of cloned genes into disease cells so that the introduced genes are expressed at suitably high levels. In principle, there are numerous different hysic-chemical and biological methods that can be used to transfer exogenous genes into human cells. The size of DNA fragments that can be transferred is in most cases comparatively very limited, and so often the transferred gene is not a conventional gene. Instead, an artificial mini gene may be used: a cDNA sequence containing the complete coding DNA sequence is engineered to be flanked by appropriate regulatory sequences for ensuring high-level expression, such as a powerful viral promote. Following gene transfer, the inserted genes may integrate into the chromosomes of the cell, or remain as extrachromosomal genetic elements (episomes).

TYPES OF GENE THERAPY

Gene therapy may be classified into two types:

(i) Germ line gene therapy.

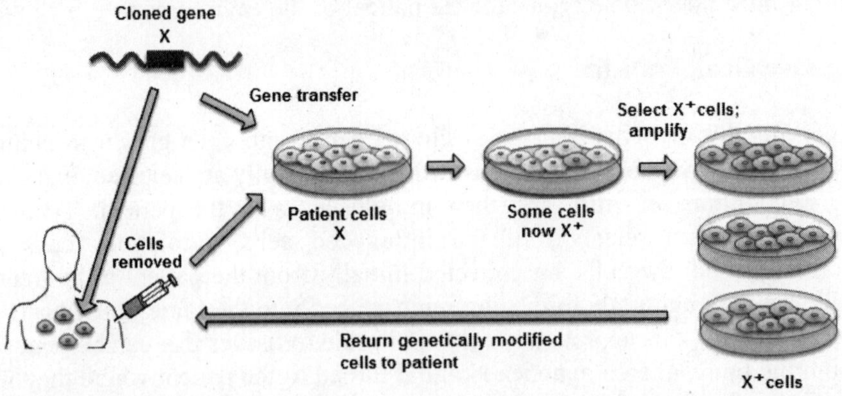

Fig. 14.1: Gene Therapy

(ii) Somatic cell gene therapy.

In case of germline gene therapy, germ cells, i.e., sperms or eggs (even zygotes), are modified by the introduction of functional genes, which are ordinarily integrated into their genomes. Therefore, the change due to therapy is heritable and passed on to later generations. This approach, theoretically, is highly effective in counteracting the genetic disorders. But, this option is not considered, at least for the present, for application in human beings for a variety of technical and ethical reasons.
In somatic cell gene therapy the gene is introduced only in somatic cells, especially of those tissues in which expression of the concerned gene is critical for health. Expression of the introduced gene relieves/eliminates symptoms of the disorder, but this effect is not heritable as it does not involve the germline(Fig. 14.2).

Somatic cell therapy is the only feasible option, and clinical trials have already started mostly for the treatment of cancer and blood disorders. This approach is divided into two groups on the basis of the end result of the process:

(i) Addition or augmentation gene therapy.

(ii) Targetted gene transfer.

Augmentation Therapy

In this type of somatic cell gene therapy, the functional gene is introduced in addition to the defective gene endogenous to the cell(s), i.e., the modified cells contain both the defective (endogenous) as well as the normal (introduced) copies of the gene. There are two general approaches to augmentation therapy.

The **first** approach was used in the first two patients on whom gene therapy was attempted to correct the genetic disorder called severe combined immune deficiency (SCID) syndrome produced by adenosine deaminase (ADA) deficiency.
Normal ADA gene copies were obtained by cloning and then.

(i) Packaged into a defective retrovirus; most of the viral genes were replaced by the ADA gene.

(ii) Lymphocytes were isolated from the patients.

(iii) The recombinant retroviruses were used to infect the lymphocytes.

(iv) Finally, the infected cells expressing the ADA gene were injected back into the patients.

The normal ADA gene was expressed in the patients, and ADA deficiency was partially corrected; this resulted in an improvement in the patient's immune system.

A variety of viral vectors have been used to deliver genes into target stem cells, e.g., lymphocytes, bone marrow cells, cultured in vitro. The stem cells themselves are obtained either from the concerned patient or from a matched donor. The reservations about safety of retroviral vectors is sought to be solved by developing suicide vectors, which cannot replicate after delivery of the gene.

Some of the different types of viruses used as gene therapy vectors:

- **Retroviruses** - A class of viruses that can create double-stranded DNA copies of their RNA genomes. These copies of its genome can be integrated into the chromosomes of host cells. Human immunodeficiency virus (HIV) is a retrovirus.

- **Adenoviruses** - A class of viruses with double-stranded DNA genomes that cause respiratory, intestinal, and eye infections in humans. The virus that causes the common cold is an adenovirus.

- **Adeno-associated viruses** - A class of small, single-stranded DNA viruses that can insert their genetic material at a specific site on chromosome 19.

Herpes simplex viruses - A class of double-stranded DNA viruses that infect a particular cell type, neurons. Herpes simplex virus type 1 is a common human pathogen that causes cold sores.

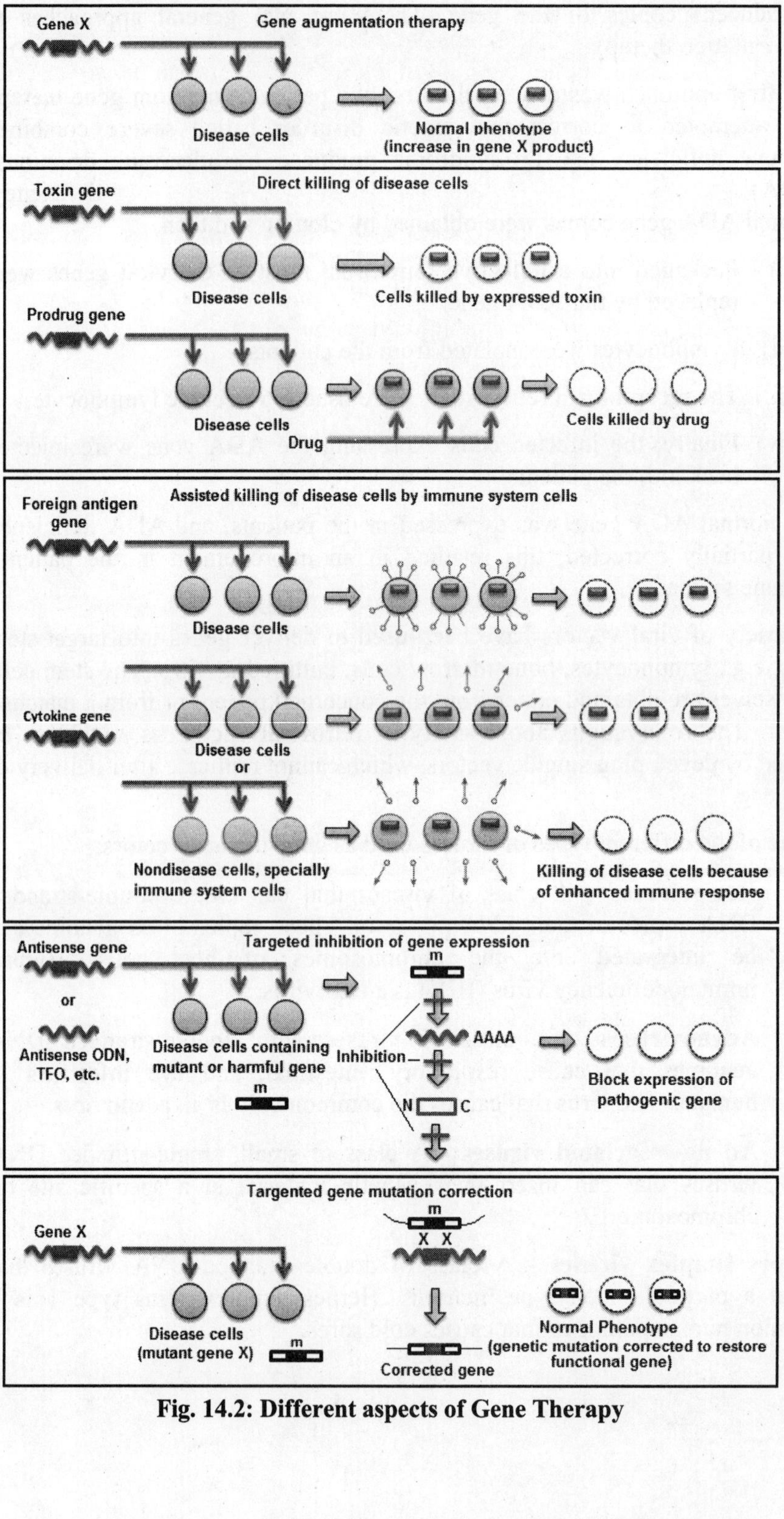

Fig. 14.2: Different aspects of Gene Therapy

Besides virus-mediated gene-delivery systems, there are several nonviral options for gene delivery. The simplest method is the direct introduction of therapeutic DNA into target cells. This approach is limited in its application because it can be used only with certain tissues and requires large amounts of DNA. Another nonviral approach involves the creation of an artificial lipid sphere with an aqueous core. This liposome, which carries the therapeutic DNA, is capable of passing the DNA through the target cell's membrane. Therapeutic DNA also can get inside target cells by chemically linking the DNA to a molecule that will bind to special cell receptors. Once bound to these receptors, the therapeutic DNA constructs are engulfed by the cell membrane and passed into the interior of the target cell. This delivery system tends to be less effective than other options.

Researchers also are experimenting with introducing a 47th (artificial human) chromosome into target cells. This chromosome would exist autonomously alongside the standard 46 --not affecting their workings or causing any mutations. It would be a large vector capable of carrying substantial amounts of genetic code, and scientists anticipate that, because of its construction and autonomy, the body's immune systems would not attack it. A problem with this potential method is the difficulty in delivering such a large molecule to the nucleus of a target cell.

The second approach is the direct injection of DNA into the tissues either as protein complexes or even as naked DNA into muscle or skin. Interestingly, these cells take up the DNA and express the gene product. Exciting results have been obtained with experimental familial hypercholesterolemia, where LDL receptor levels have been augmented by injection of the gene as a sialoglyco protein complex.

Targetted Gene Transfer

Targetted gene transfer or gene targetting uses homologous recombination to replace the endogenous' gene with the functional introduced gene. The first case of such gene transfer (in 1985) was used to disrupt the human 13-globin gene in cultured cells. Subsequently, over 100 mammalian genes have been modified by this approach.

Gene targetting can be used either to inactivate (by disruption) a functional endogenous gene or to correct a defective one. The vectors employed for gene targetting are of two types:

(i) Insertion vectors.

(ii) Replacement vectors

The insertion vector is linearized by restriction cleavage within the sequence to be targetted; the targetted sequence provides the site for recombination and is different from the gene to be introduced. Hence the sequence to be introduced is located in the inner region of the vector and is flanked by the sequences involved in recombination. A recombination of such a vector with its homologous cellular sequences produces a duplication of the targetted sequence; this is called Insertional recombination.

In contrast, a linearized replacement vector has the two halves of the target gene at its twp ends. Recombination occurs within the two halves of the target gene, replacing a portion of the endogenous gene sequence by that of the introduced gene: this is called replacement recombination. There is no duplication of sequences, and the target gene becomes disrupted.

A strategy has been devised to modify only a small sequence of the target gene without the attendant gene duplication/disruption produced by insertional/replacement recombination. This approach, called in out method of gene targetting, consists of the following two steps:

1) The first step called "in" step, is targetted gene transfer using an insertion vector; the appropriately targetted cell will have a gene duplication.

2) The second step, termed as "out" step, depends on either intrachromosomal recombination (between the introduced and the endogenous genes) or unequal sister chromatid exchange between homologous chromosomes. The recombination product of interest is a chromosome, which has only a single and functional copy of the introduced gene.

Gene targetting is the strategy of choice for gene therapy for the following reasons:

1) The targetted gene is changed in a precise and specific manner.
2) The introduced functional gene is placed in the same context, i.e., is flanked by the same DNA sequences, as the replaced endogenous gene.
3) No other gene of the genome is affected. The major limitation of the approach is the low frequency of homologous recombination, but it is expected that targetted therapy would become a feasibility for many genetic diseases in the near future.

15

Gene Tagging and Gene Knockout Technology

TRANSGENIC TECHNOLOGY

During the last few years, new methods have been developed which, unlike conventional breeding techniques, allow the genetic composition of organisms to be modified directly and selectively rather than indirectly by estimating the breeding values on the basis of phenotypes in order to establish ranking order for selection. In principle, these new techniques of genetic manipulation can be categorized into those which treat the entire genome as a compact unit (genome manipulation; e.g. nuclear transfer) and those in which individual genes are manipulated (gene manipulations; e.g. gene transfer). Further, these techniques can be differentiated into those allowing manipulation of somatic cells (e.g. somatic gene therapy) and those directed at altering the germ line of animals. The latter techniques give rise to transgenic lines of animals characterized by the stable transmission of the genetic modification.

Transgenic technology refers to the genetic engineering of an organism so that its genomic DNA is altered to over- or under express certain genes. All progeny of a transgenic organism will share the same genotype as the parents (assuming both parents are homozygous). A knockout refers to a transgenic organism in which a gene has been replaced or disrupted with an artificial piece of DNA. The most common type of transgenic organism used in research is the knockout mouse, though knockout rats and knockout rabbits have also been developed. It has become routine to develop knockout mice with disruptions in specific genes. These knockout mice allow scientists to view the effect of this gene disruption on the resulting phenotype. Often the phenotype is a direct result of the gene knockout and can provide clues as to the biological role of the gene, but occasionally the phenotype can also be the result of compensatory or indirect effects of the gene knockout. Sometimes the procedure to create a transgenic organism can result in a phenotype completely unrelated to the disrupted gene, but is an effect of the artificial DNA used to create the disruption. In addition, some gene knockouts create a lethal phenotype where the organism fails to develop in utero, making in vivo studies exceedingly difficult.

The application of gene transfer techniques has furnished new insights in the developmental biology and the principle underlying tissue-specific gene expression. Gene transfer allows the development of new production system for pharmaceutically important proteins. As far as animal production is concerned, gene transfer appears to be a promising technique for improving disease resistance, performance of the quality of animal products by modifying, for example, metabolic pathways, and hormone status.

The significance of transgenic organisms can be depicted by two most common reasons:

1) Some transgenic animals are produced for specific economic traits. For example, transgenic cattle were created to produce milk containing particular human proteins, which may help in the treatment of human emphysema.
2) Other transgenic animals are produced as disease models (animals genetically manipulated to exhibit disease symptoms so that effective treatment can be studied). For example, Harvard scientists made a major scientific breakthrough when they received a U.S. patent for a genetically engineered mouse, called Onco Mouse or the Harvard mouse, carrying a gene that promotes the development of various human cancers.

TRANSGENIC PROCEDURES

The ability to insert an exogenous (or foreign) gene into the mouse genome by direct injection into the pronuclei of zygotes was achieved just two decades ago. The term *transgenic* was applied to mice expressing exogenous DNA that had been produced using this technique. With this method, the gene of interest is inserted into a random locus in the mouse genome, and is expressed ''in trans,'' i.e., not in its usual genetic locus. The techniques required for introducing transgenes into the mouse genome have been highly refined, permitting their widespread use. Since the development of this technique, many thousands of lines of transgenic mice have been generated, and it has been the most widely utilized technique of genetic manipulation in mice.

Production of Transgenic Organisms

Until recently, selective breeding was the only way to enhance the genetic features of domesticated animals. However the combination of the successful transfer of genes into mammalian cells and the possibility of creating genetically identical animals by transplanting nuclei from embryonic tissue into enucleated eggs (nuclear transfer, nuclear cloning) led researchers to consider putting single, functional genes or gene clusters into the chromosomal DNA of higher organisms. Conceptually, the strategy used to achieve this end is simple. A cloned gene is injected into the nucleus of a fertilized egg. The inoculated fertilized eggs are implanted into a receptive female (because successful completion of mammalian embryonic develop outside of a female). Some of the offspring derived from the implanted eggs carry the cloned gene in all of their cells. Animals with the cloned gene integrated in their germ-line cells are bred to establish new genetic lines.

The genetic improvement of multicellular organisms by the introduction of relevant transgenes is only slowly being realized. However, transgenesis has become a powerful technique for studying fundamental problems of mammalian gene expression and development, for establishing animal model systems for human diseases, and for using the mammary pharmaceutically important proteins in milk. With this application in mind, the term **pharming** was coined to convey the idea that milk from transgenic farm (**Pharm**) animals can be source of authentic human protein drugs or pharmaceuticals.

Transgenic Mice

Mouse is the most preferred mammal for studies on gene transfers due to its many favourable features like short oestrous cycle and gestation period, relatively short generation time, production of several offsprings per pregnancy, convenient IVF, successful culture of embryos *in vitro* at least for a particular period of time, production and maintenance of ES cell lines, availability of a diverse array of genetic stocks, etc.

Transgenic technology has been developed and perfected in the laboratory mouse. Since the early 1980s, hundreds of different genes have been introduced into various mouse strains. These studies have contributed to an understanding of gene regulation, tumour development, immunological genetics of development, and other biological process of fundamental interest. Transgenic mice have also played a role in examining the feasibility of the industrial production of human therapeutic drugs by domesticated animals and in the creation of transgenic strains that act as biomedical models for various human genetic diseases. For transgenesis, DNA can be introduced into mice by one of the following methods:

1) Retroviral vectors that infect the cells of an early-stage embryo prior to implantation into a receptive female.
2) Microinjection into the enlarged sperm nucleus (the male pronucleus) of a fertilized egg.
3) Introduction of genetically engineered embryonic stem cells into an early-stage developing embryo before implantation into a receptive female. Embryonic stem cells (ES cells) are harvested from the inner

cell mass (IGM) of mouse blastocysts. They can be grown in culture and retain their full potential to produce all the cells of the mature animal, including its gametes.

Methods of Production of Transgenic Mice

Techniques for producing transgenic mice involve the microinjection of DNA constructs into fertilized mouse eggs. DNA constructs used for the generation of transgenic mice typically consist of a gene of interest located 3' to promoter sequences selected to produce a desired distribution of gene expression. The maximum length of the DNA sequence that may be successfully incorporated into the mouse genome is not known, and up to 70 kilobase (kb) DNA fragments have been successfully integrated. The transgene is linearized and purified from prokaryotic vector sequences. For optimal integration efficiency, about 1 to 2 picoliter (pL) of DNA at a concentration of 1 to 2 ng/μL (corresponding to a few hundred molecules of a 5-kb DNA fragment) is microinjected into the male pronucleus of a fertilized mouse egg. Although labor intensive, direct injection of DNA into the pronucleus results in much higher rates of integration of transgenes than other known methods of transformation. After microinjection, the embryos are surgically transferred into the oviduct of pseudopregnant mice. Pseudopregnant females are generated by matings with vasectomized males. The act of copulation initiates the endocrine changes of pregnancy, providing a suitable uterine environment for the survival and implantation of the transferred embryos. The foster mothers give birth 19 to 21 days after oviduct transfer. For genotyping, DNA is typically isolated from mouse tail biopsies and screened for the presence of the transgene by Southern blotting or polymerase chain reaction (PCR). Typically, about 20% to 40% of the mice that develop to term possess the transgene. In the majority of cases, integration of the transgene occurs during the one-cell stage, so that the transgene is present in every cell of the transgenic mouse (Fig. 15.1). Integration usually occurs at a single random chromosomal location, and, for reasons that are not fully understood, there are usually multiple copies of the transgene inserted as head-to-tail concatamers. Mice identified to possess the integrated transgene are referred to as *founders*. The founders are typically used in a breeding strategy to produce animals that are homozygous for the transgene insertion.

Gene Targeting in Mice

Gene targeting is the insertion of DNA into a specific chromosomal location; it is often used to inactivate a specific gene in the, genome. Gene inactivation is a common means of elucidating how a specific gene functions. In gene targeting, a cloning vector (called a targeting vector), which harbours the gene to be inserted, recombines with a region of the target chromosome that is homologous with a DNA region on the targeting vector(Fig. 15.1). This process is called as homologous recombination, and may be defined as the exchange of identical segments between two DNA molecules that have identical or almost identical sequences. In yeast, gene integration occurs ordinarily by homologous recombination. But in mammals, random DNA integration is far more frequent than homologous recombination. The frequency of integration by homologous recombination appears to be only 0.1 to 1.0% of random integration events; this makes the recovery of transgenics representing targeted gene transfer quite difficult. Recent refinements in the techniques, however, have increased this frequency to 10% or even 50% of random integrations.

The approaches for the identification of homologous recombinations are as follows:

1) When inactivation or activation of the test gene occurs due to homologous recombination, but not by random integration, and produces a selectable phenotype.
2) Alternatively, a large number of transfected cells/clones may be screened, often using PCR amplification, to identify those having homologous recombination.

A selectable marker gene (for example, bacterial neomycin phosphotransferase, which confers resistance to the antibiotics neomycin and kanamycin) ensures that the desired recombination event is selected. Antibiotics included in the growth medium of cells transformed with the targeting vector and inserted gene allows only those cells that have the correct insertion into the chromosome to survive.

Knock-out and Knock-in Technology

In order to study the relationship between proteins and gene function, scientists now can prevent the manufacturing of the protein by a specific gene. By disabling the gene from a test organism, and then producing descendents that contain the copies of the disabled gene, it is possible to observe the descendents' development in the absence of a particular protein. This practice, referred to as knock-out technology, is attempted to shut down or turn off a particular gene. Therefore, the mouse has been the mammal in which knock-out technology has been most generally applied. In essence, a "knock-out" organism is created when an ES cell is genetically engineered and then inserted into a developing embryo (Fig. 15.2). The embryo is then inserted surgically into the womb of the host (e.g. a fern mouse). Once the embryo has matured, a portion of its stem cells will produce egg and sperm with the knocked-out gene. A gene can also be altered in function, in contrast to being deleted.

When a gene is altered but not shut down, a "targeted mutation" effect is created. This practice is referred to as knock-in technology, whereby a life form has an altered gene "knocked" into it. Gene knock-out/ knock-in

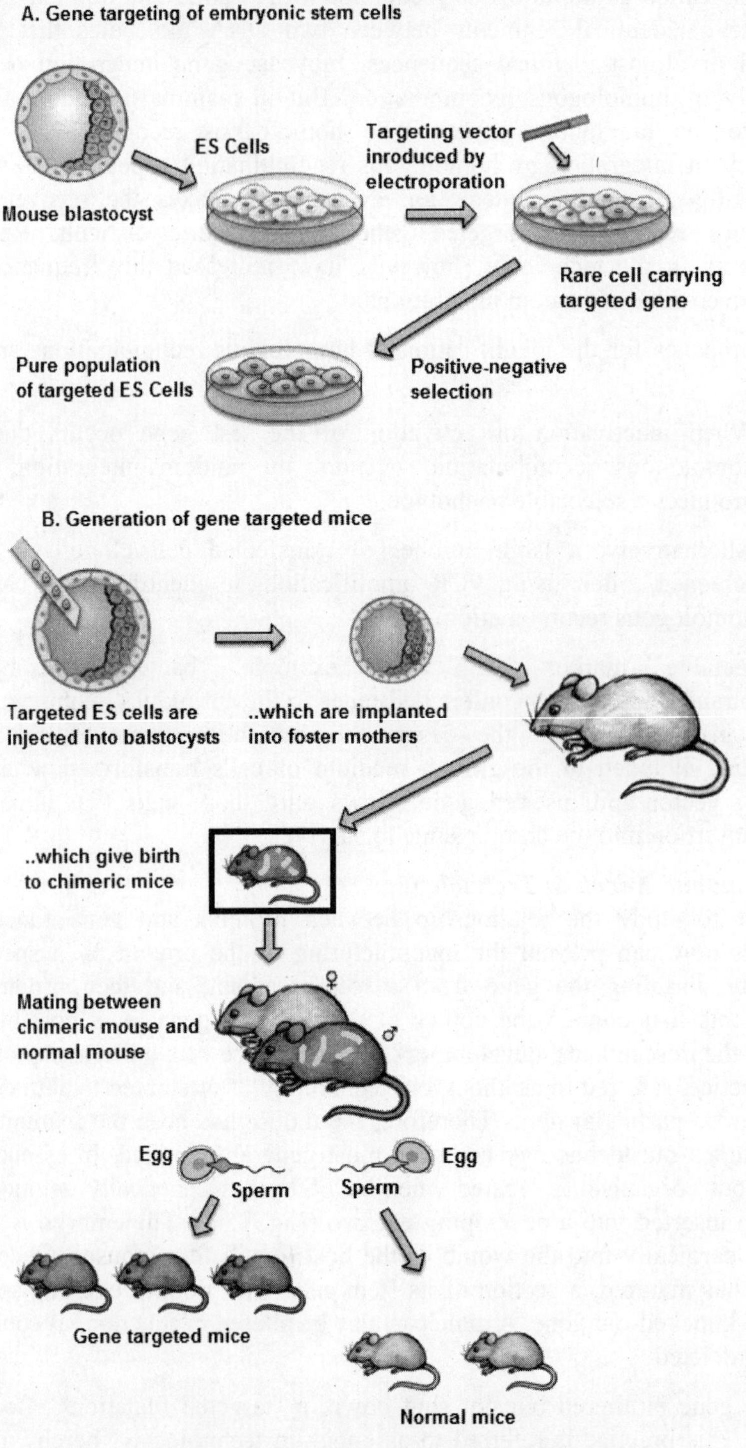

Fig. 15.1: Transgenic Mouse production

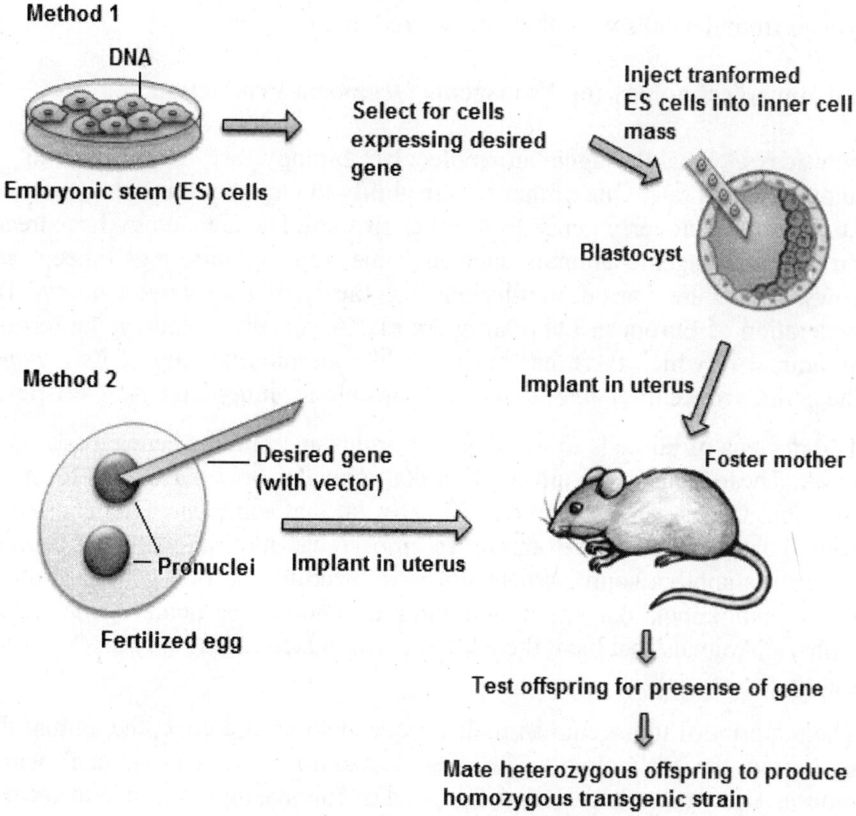

Fig. 15.2: "Knock out" Mice

technology is well-established as experimental tool in mice, due to the availability of ES cell lines. The principle is to take advantage of a rather rare event that occurs after introduction of DNA into cells - homologous recombination between identical sequence in the genome and the transfecting DNA. In the most common protocol, a selectable marker such as the neomycin-resistance gene is inserted within a piece of DNA corresponding to the portion of a gene of interest.

After transfection of cells by this construct and selection for the marker (by growth in medium containing the neomycin-related antibiotic G418) the selected cells are screened to identify the small fraction that has one copy of the gene of interest disrupted by the marker. Progeny animals derived from the cells will be heterozygous for the "knocked – out" gene; breeding to obtain straightforward. Because the process is so inefficient, very large numbers of transfected cells was to be screened, making the use of cultured cells essential, since, it would be impracticable to screen large numbers of progeny from microinjected eggs. The galactosyl transferase- knock-out pigs were generated from cultured foetal fibroblasts manipulated in this way. Nuclei from the cells were then transferred into oocytes.

Advance Technology for Transgenic Organism Production

Nowadays, breakthroughs in molecular biology are happening at an unprecedented rate. One of them is the ability to engineer transgenic animals, i.e., animals that carry genes from other species. The technology has already produced transgenic animals such as mice, rats, rabbits, pigs, sheep, and cows. There are various definitions for the term *transgenic animal.* The Federation of European Laboratory Animal Associations defines the term as an animal in which there has been a deliberate modification of its *genome*, the genetic makeup of an organism responsible for inherited characteristics.

The nucleus of all cells in every living organism contains genes made up of DNA. These genes store information that regulates how our bodies form and function. Genes can be altered artificially, so that some characteristics of an animal are changed. For example, an embryo can have an extra, functioning gene from another source artificially introduced into it, or a gene introduced which can knock out the functioning of another particular gene in the embryo. Animals that have their DNA manipulated in this way are knows as transgenic animals.

The majority of transgenic animals produced so far are mice, the animal that pioneered the technology. The first successful transgenic animal was a mouse. A few years later, it was followed by rabbits, pigs, sheep, and cattle.

Scientists can now produce transgenic animals because, since Watson and Crick's discovery, there have been breakthroughs in:

- Recombinant DNA (artificially-produced DNA).
- Genetic cloning.
- Analysis of gene expression (the process by which a gene gives rise to a protein).
- Genomic mapping

The underlying principle in the production of transgenic animals is the introduction of a foreign gene or genes into an animal (the inserted genes are called transgenes). The foreign genes "must be transmitted through the germ line, so that every cell, including germ cells, of the animal contain the same modified genetic material." (*Germ cells* are cells whose function is to transmit genes to an organism's offspring).

To date, there are three basic methods of producing transgenic animals:

- DNA microinjection.
- Retrovirus-mediated gene transfer.
- Embryonic stem cell-mediated gene transfer

Gene transfer by microinjection is the predominant method used to produce transgenic farm animals. Since the insertion of DNA results in a random process, transgenic animals are mated to ensure that their offspring acquire the desired transgene. However, the success rate of producing transgenic animals individually by these methods is very low and it may be more efficient to use cloning techniques to increase their numbers. For example, gene transfer studies revealed that only 0.6% of transgenic pigs were born with a desired gene after 7,000 eggs were injected with a specific transgene.

Contribution of transgenic animals to human welfare

The benefits of these animals to human welfare can be grouped into areas:

1) Agriculture.
2) Medicine

The examples below are not intended to be complete but only to provide a sampling of the benefits.

Agricultural Applications

Breeding: Farmers have always used selective breeding to produce animals that exhibit desired traits (e.g., increased milk production, high growth rate). Traditional breeding is a time-consuming, difficult task. When technology using molecular biology was developed, it became possible to develop traits in animals in a shorter time and with more precision. In addition, it offers the farmer an easy way to increase yields.

Quality: Transgenic cows exist that produce more milk or milk with less lactose or cholesterol, pigs and cattle that have more meat on them, and sheep that grow more wool. In the past, farmers used growth hormones to spur the development of animals but this technique was problematic, especially since residue of the hormones remained in the animal product.

Disease resistance: Scientists are attempting to produce disease-resistant animals, such as influenza-resistant pigs, but a very limited number of genes are currently known to be responsible for resistance to diseases in farm animals.

Medical Applications

Xenotransplantation: Patients die every year for lack of a replacement heart, liver, or kidney. For example, about 5,000 organs are needed each year in the

United Kingdom alone. Transgenic pigs may provide the transplant organs needed to alleviate the shortfall. Currently, xenotransplantation is hampered by a pig protein that can cause donor rejection but research is underway to remove the pig protein and replace it with a human protein.

Nutritional supplements and pharmaceuticals: Products such as insulin, growth hormone, and blood anti-clotting factors may soon be or have already been obtained from the milk of transgenic cows, sheep, or goats. Research is also underway to manufacture milk through transgenesis for treatment of debilitating diseases such as phenylketonuria (PKU), hereditary emphysema, and cystic fibrosis. In 1997, the first transgenic cow, Rosie, produced human protein-enriched milk at 2.4 grams per litre. This transgenic milk is a more nutritionally balanced product than natural bovine milk and could be given to babies or the elderly with special nutritional or digestive needs. Rosie's milk contains the human gene alpha-lactalbumin.

Human gene therapy: Human gene therapy involves adding a normal copy of a gene (transgene) to the genome of a person carrying defective copies of the gene. The potential for treatments for the 5,000 named genetic diseases is huge and transgenic animals could play a role. For example, the A. I. Virtanen Institute in Finland produced a calf with a gene that makes the substance that promotes the growth of red cells in humans.

T-DNA AND TRANSPOSON TAGGING

Gene Isolation by Tagging

In the last few years the complete sequence of genomes of several important model eukaryotic species have been published, most recently the human genome itself. With the wealth of information that has been generated by these genome projects, the next important step is to find out what all the newly discovered genes actually do. This is the burgeoning field of functional genomics which aims to determine the function of all transcribed sequences in the genome and all the proteins that are made.

Isolation of genes by conventional techniques require knowledge of a gene product. In routine procedures for gene cloning, genomics or expression libraries are screened with probes made of mRNA or protein respectively. Also, based on the cDNA or protein sequence, oligonucleotide primers can be designed to apply a PCR approach. But for the genes of unknown products, alternative stratigies that have a genetic basis are required like map based cloning approaches. However, such approaches do not involve mutant phenotypes that may hamper the identification of genes. Microarrays and variants of yeast two hybrid system are some techniques of functional genomics. However, perhaps the most important way to establish a gene's function is to see what happens when the gene is either mutated orinappropriatley expressed in context of the whole organism. Such techniques allow the high throughput analysis of gene function, which can be the only way we can begin to understand what 1000-40000 genes in genomes of higher eukaryotic cells are for. Gene tagging is one such technique, others being: insertional mutagenesis and entrapment constructs. During the last two decades significant progress has been made in the techniques for isolation of a variety of genes, including those for:

(i) Ribosomal RNA.

(ii) Specific protein products.

(iii) Phenotypic traits with unknown product.

(iv) Regulatory functions e.g. promoter genes, etc.

Different techniques have been used for the isolation of these different types of genes.

In some cases of isolation of a gene, the interested gene is whose phenotypic effect is known, but the gene product has not been identified or can not be isolated. Such genes include those for morphological traits like dormancy, photoperiodicity, disease resistance, etc. This area of research, in which genetics is studied by isolating the gene first without knowing the gene product, is often described as reverse genetics. The methods used for the isolation of these genes are different from those used for genes coding for known proteins.

Gene Isolation by Transposon Tagging

Transposons are sequences of DNA that can move around to different positions within the genome of a single cell, a process called transposition. In the process, they can cause mutations and change the amount of DNA in the genome. Transposons were also once called jumping genes, and are examples of mobile genetic elements. They were discovered by Barbara McClintock early in her career for which she was awarded a Nobel prize in 1983.

There are a variety of mobile genetic elements, and they can be grouped based on their mechanism of transposition. Class I mobile genetic elements, or retrotransposons, copy themselves by first being transcribed to RNA, then reverse transcribed back to DNA by reverse transcriptase, and then being inserted at another position in the genome. Class II mobile genetic elements move directly from one position to another using a transposase to "cut and paste" them within the genome. Transposons make up a large fraction of genome sizes which is evident through the C-values of eukaryotic species. The sheer volume of seemingly useless material initially puzzled researchers, so that it was termed "Junk DNA" until further research revealed the critical role that it played in the development of an organism. They are very useful to researchers as a means to alter DNA inside a living organism.

The molecular isolation of transposable elements now permits the cloning of genes in which the element resides. The major advantage of this system is that genes whose function is not known can be cloned. The first step in this procedure is to identify a plant stock that is mutant for a specific trait because a transposable element has been inserted into and inactivated the gene. Next, a genomic library (often in bacteriophage lambda) of the plant stock is created. This library is then screened with a clone for the transposable element. Any clone that is selected from the screening will contain the element. In the clone, sequences for the mutated gene will lie adjacent to the element. A sublcone containing sequences from the gene is then developed from the non-transposable element DNA of the original clone. This clone is then used to screen a genomic library containing DNA from a normal plant. In this manner, any clone that is selected should contain a full, normal copy of the gene.

Because this system is so powerful, scientist has begun introducing elements from corn and *Antirrhinum* into other species using transformation techniques. It has been demonstrated that these elements can be induced to move from one location to another in the new species. If this movement is coupled with the appearance of mutant phenotype, then the gene responsible for the phenotype can cloned in that particular species. These techniques have now allowed the use of transposon tagging in plant species in which active transposable elements have not been identified (Fig. 15.3).

Gene Isolation by T-DNA Tagging

Agrobacterium tumefaciens is a soil borne bacterium. Like Rhizobium, the nitrogen fixing bacterial genus, Agrobacterium has developed a way of living in and deriving nourishment from plant tissues. However, unlike Rhizobium, *Agrobacterium* is a parasite and provides no benefit to the plant that it colonizes. Instead, it causes crown gall disease. *Agrobacterium* resides in its large extrachromosomal plasmid, subsequently named the tumor inducing plasmid (Ti plasmid). A section of Ti plasmid (T -region) was transferred to and maintained as T-DNA in the transformed plant cell by insertion into the plant nuclear genome. *Agrobacterium* plant cell transformation requires wounding the cell. The virulence genes borne on the Ti plasmid are activated by certain phenolic compounds released by injured plant cells, which control excision of the T-DNA portion from the Ti plasmid of the bacterium. The, T-DNA becomes integrated into the host chromosomal DNA, thereby transforming the host plant genetically to form tumors, which are hormone independent.

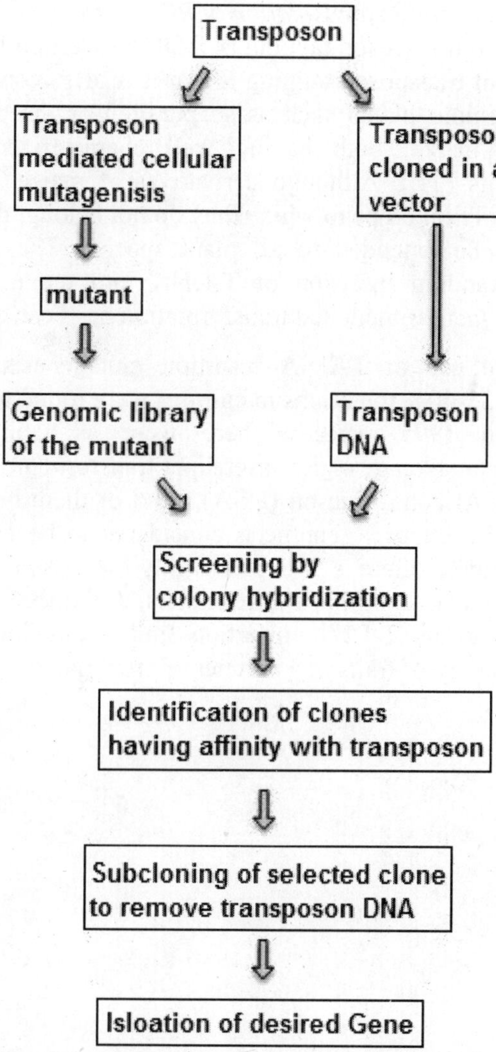

Fig. 15.3: Transposon tagging

T-DNA

T -DNA is the transferred DNA or transforming DNA of the Ti plasmid. The first evidence indicating that Ti plasmids were involved in a gene transfer system resulted from a genetic analysis of Ti plasmid functions. Direct evidence for such a transfer was obtained first by renaturation kinetics analysis. A more precise picture resulted from DNA/DNA hybridization using the Southern gel blotting technique. Southern blotting analysis of octopine and nopaline tumor lines demonstrated that a well-defined part of both Ti plasmids is transferred and stably maintained in plant cells.

T-DNA Insertion Mutagenesis for Isolation of Plant Genes

A technique similar to transposon tagging is T-DNA insertion for isolation of genes. A limitation of transposon tagging is that it is often confined to either the self fertilized diploid plants such as snapdragon or the cross-fertilized diploid crops like maize, both having well characterized endogenous transposable elements (TE). Although derivatives of maize Ac have been successfully used as TE in crops to which they do not belong, their use is still limited and cannot be extended to all plant species. Therefore, T-DNA tagging involving random insertion of T-DNA into the genome, during *Agrobacterium tumefaciens* mediated transformation has been exploited.

The most extensive use of T-DNA insertion mutagenesis has been in *Arabidopsis*, where 35-40% mutations in cultures were found to be tagged by T-DNA inserts. In 1995, already there were as many as 14,000 Transformants, with an average of 1.5 inserts per transformant were available at the University of Arizona, Tucson (USA), 40% of them being due to T-DNA insertion. If *Arabidopsis* genome is considered to be 130 Mb and T-DNA insertion is random, there is 50% probability for any individual gene to be tagged among the above 14,000 transformants. A number of these genes have been cloned using T-DNA insertion and are being utilized for improvement of a variety of traits in a number of crop plants.

References

http://www.pearsoned.co.in

http://www.Molecular-Plant-Biotechnology.info

http://www.ambion.com

http://www.studentsguide.in

http://www.scq.ubc.ca/the-yeast-two-hybrid-assay-an-exercise-in-experimental-eloquence/

http://www.nobelprize.org/

http://www2.wmin.ac.uk/~redwayk/lectures/sdm.htm

http://www.ncbi.nlm.nih.gov/books/NBK7571/

http://www.koreanbio.org

http://www.nature.com/nprot/journal/v3/n6/fig_tab/nprot.2008.67_F1.html

http://www.piercenet.com

http://www.nationaldiagnostics.com

http://www.biotechnet-waldshut.de/boston_lessons/material

http://www.fao.org/ag/aga/agap/frg/feedback/war/u5700b/u5700b04.htm

http://www.fastol.com/~renkwitz/microarray_chips.htm

http://gi.cebitec.uni-bielefeld.de/comet/chiplayout/index.html

http://www.ncbi.nlm.nih.gov/books/NBK7569/

http://www.nobelprize.org/nobel_prizes/medicine/laureates/2007/adv.html

http://www.wikipedia.org/

Index